systematic commercial refrigeration service

Raymond K. Schneider

Business News Publishing Company
Birmingham, Michigan

Library of Congress Cataloging in Publication Data
Schneider, Raymond K.
Systematic commercial refrigeration
service.

Includes index.
1. Refrigeration and refrigerating
machinery—Maintenance and repair.
I. Title
TP492.S38 621.5 78-14464
ISBN 0-912524-18-9

Printed in the United States of America

Dedication

To Mom and Dad for their continued encouragement in all my undertakings.

Acknowledgment

My thanks to Virginia Teska for her clerical help, that eased a major task, and to the many companies that responded to my request for pictorial material to complement the text.

Contents

Introduction

The refrigeration field has often been considered to be one entity; that is, one broad category for the purpose of describing employment possibilities or business volume or labor statistics. Talking to someone not involved in the field is likely to bring a statement such as, "Oh, so you're in the refrigeration business. I heard that's a pretty good job," and that seems to say it all. To someone in the business, however, such a statement just begins to tell the story. Just where in that vast field does this man's job lie? Here is just a short list:

- Refrigeration
  - Sales
  - Service
  - Installation
    - Domestic
    - Commercial
    - Industrial
    - Process
    - Institutional

We could look under a more powerful microscope and might find that the man is a sheetmetal worker, a steamfitter, a parts man, a dispatcher, a truck driver, a helper, a journeyman, etc., etc. Before too much effort was expended, we would have a list, several pages long, of the types of jobs that exist under the broad classification, Refrigeration.

In the following pages, we will bite off a part of this vast field and look closely at commercial refrigeration. Commercial refrigeration means different things to different people. Throughout this book we shall take it to mean refrigeration equipment used in establishments dealing with the general public. In this category, we find food stores, restaurants, taverns, cafeterias, fast-food stands, beverage stores, fish stores, supermarkets and other such places of business.

We will look at the refrigeration equipment used in these establishments. We'll see how it operates, how it can malfunction and, even more important, how it can be repaired. We will also look at the type of evaluation that went into the selection of a particular piece of equipment.

After you've been in the business a while, you'll find out that service problems break down something like this:

1. Refrigerant leaks—30%
2. Electrical problems—43%
3. Mechanical problems—15%
4. Other problems—12%

It is unlikely that you will do any significant amount of service work without encountering electrical problems. Since a substantial part of the commercial refrigeration service business is electrical, we shall spend a good deal of effort in describing electrical troubleshooting procedures.

Mastery of categories 1 and 3, leaks and mechanical, will come with a few months experience. Therefore, we'll concentrate on category 4, the 12% of the business that will bring on gray hairs and an odd knot in the pit of your stomach.

Now, it's time to move along and look at commercial refrigeration. We'll see the types of equipment to be encountered in this important field, look at their method of operation, and spend some time on refrigeration circuit and electric circuit troubleshooting.

1

Reach-in Refrigerators

We are all familiar with the refrigerator/freezers we have in our homes. Both a refrigerator section, with temperatures of about 35 to 40°F, and a freezer section, with temperatures of about —5 to 0°F, are contained in one appliance. While there are probably more of these refrigeration units than any other type, they are really not the simplest of all devices. The design is, in fact, rather complex since two different temperatures are to be maintained while utilizing only one compressor unit. The fact that this is done without accessories, such as evaporator pressure regulators or thermostatic expansion valves or other sophisticated controls, is what makes the system somewhat complex.

Commercial refrigerators and freezers differ from the domestic versions in several ways. In a restaurant or market, such a box may be opened and closed several *hundred* times a day by people with varying concerns for the life of the unit, so you appreciate some of these differences:

1. They are stronger, to withstand heavier usage.
2. They have fewer plastic parts.
3. Stainless steel and/or aluminum are more widely used because painted steel is subject to scratching.
4. Larger compressor units are used because of higher cooling loads due to more frequent door openings.
5. Larger size boxes are available for different applications.

While many domestic boxes are to be found in smaller commercial establishments, a far larger number of reach-in refrigerators and freezers, doing the job for which they were designed, will be employed.

We'll look first at the reach-in refrigerator unit's various configurations and then deal with reach-in freezers. The main differences between refrigerators and freezers are the low side operating pressures (freezers having low pressures) and in the need for a means of artificially defrosting freezers. For the most part, everything said about refrigerators will be quite applicable to freezers.

REACH-IN REFRIGERATORS

There are probably more reach-in refrigerators than any other type of commercial refrigeration equipment. It is the basic piece of equipment for any food handling enterprise. Generally, we can talk of two types of reach-in: that is, the storage type, Figure 1-1, and the display type, Figure 1-2. The display box differs from the storage box in two main areas: first, one or more doors will be of glass to permit customers to see the product and, second, since the doors are glass, with somewhat reduced insulating capability, the refrigeration system will usually have a greater capacity to overcome the increased heat leakage. Other than these differences, the cabinets are identical in construction, component configuration, and wiring.

Reach-in refrigerators are generally used for short-term storage of perishable food items. Box temperatures on the order of 35° to 45°F are typical.

FIG. 1-1. Three-door, storage-type, reach-in refrigerator. *Courtesy Victory Mfg. Div., McGraw-Edison Co.*

Walk-in refrigerators are usually employed for long-term storage, but reach-in boxes can also be used. Temperatures of 33° to 36°F are usually held for long-term storage.

The main reason for the popularity of reach-in refrigerators is their convenient size and ease of access. On the basis of cost per cubic foot of storage space, the walk-in is generally less expensive than the reach-in, but more cumbersome and inconvenient to use, particularly in a food service application.

FIG. 1-2. Double sliding door, display-type, reach-in refrigerator. *Courtesy Victory Mfg. Div., McGraw-Edison Co.*

CABINET CONFIGURATION

Once we've specified a display box or a storage box, the next option is whether the box is to be self-contained or remote. That is, is the condensing unit to be mounted within the box itself or at some remote location with tubing run to the box? If self-contained, the condensing unit is either mounted below the storage compartment, Figure 1-3, or above it, Figure 1-4. There is something to be said for either location. If the condensing unit is mounted at floor level, it may tend to pick up dirt rather quickly. This dirt is drawn in by the condenser fan, lodges in the condenser coil and retards the flow of air over the coil surface. This can

FIG. 1-3. Self-contained condensing unit below the left door occupies usable space, tends to pick up dirt. *Courtesy Foster Refrigerator Corp.*

lead to all sorts of problems as we'll see. Being low, however, it does tend to have the coolest air in the room passing over it. The top mounted unit usually will not get dirty as quickly, being off the floor. However, since it is 6′ to 7′ off the floor, the air passing over the condenser can be quite warm compared to air at floor level. The answer to these perplexing alternatives may very well be the remote condenser which can be placed in a clean, cool area and solve both problems. Don't be surprised, however, if this remote unit has been placed in a hot, dirty area because, more often than not, the main consideration in placing this unit is to just get it out of the way.

FIG. 1-4. Top-mounted condenser is cleaner and has more usable space, but is cooled with warmer air. *Courtesy Foster Refrigerator Corp.*

CONDENSING UNITS

A condensing unit, Figure 1-5, is comprised of a compressor and condenser mounted on a common base. Condensing units employed with thermostatic expansion valves usually have a receiver mounted on the base while capillary tube units do not have a receiver. Generally, air cooled condensers are used although, in remote applications, there is no reason why a water cooled unit couldn't be used.

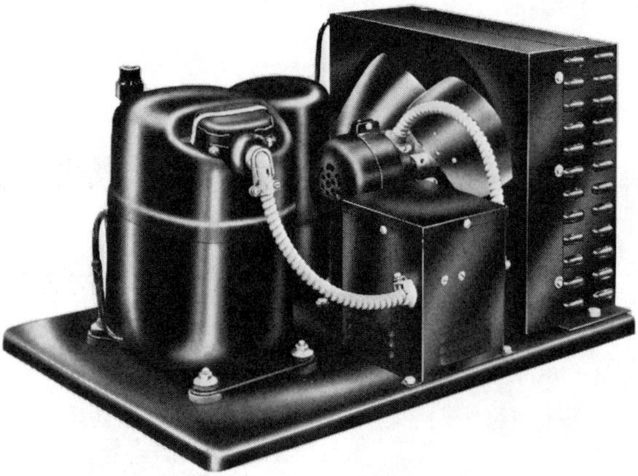

FIG. 1-5. The top condensing unit is employed with a TXV system. Note the receiver to the right of the compressor. The lower condensing unit is employed with capillary tube systems. *Courtesy Tecumseh Products Co.*

EVAPORATORS

Evaporators, Figures 1-6 and 1-7, are usually mounted inside the storage compartment. Both draw-through and blow-through fans are used in the sloped front, semi-circular, flat, and oval configurations.

Mullion blower coils, Figure 1-8, are used to increase the usable space in the storage compartment. They give a neater inside appearance and are a useful sales feature. However, they are a bit more difficult to service because the contents of the box must be removed to get at the coil. A coil similar to a mullion coil can be mounted across the top of the box, again resulting in a clean appearance.

FIG. 1-6. Wilmington evaporator blower coil typical of those found in reach-in refrigerators. *Courtesy Climate Control Div., Singer Corp.*

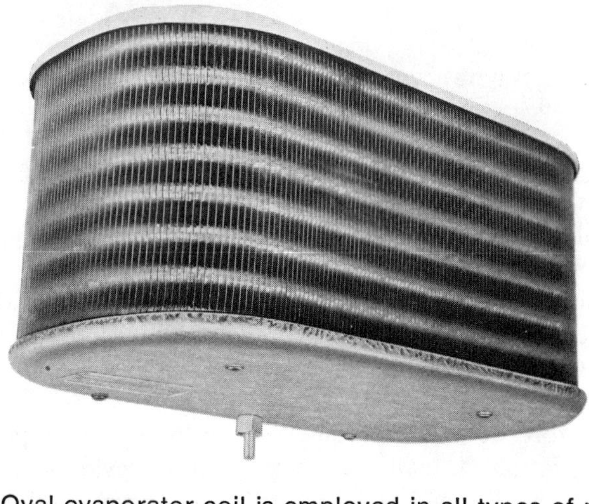

FIG. 1-7. Oval evaporator coil is employed in all types of refrigerators. *Courtesy Bohn Heat Transfer Div., Gulf + Western Mfg. Co.*

FIG. 1-8. Mullion-type evaporator blower coil. *Courtesy Bohn Heat Transfer Div., Gulf + Western Mfg. Co.*

One configuration maximizes storage space by having the evaporator mounted on top of the box, outside the storage space. Air passages are used to circulate the air from the compartment over the evaporator coil and then back to the compartment.

There should be no difference in the servicing of any of these boxes. Some may be easier to work on than others, but they all have an evaporator (usually with a fan), a condensing unit, controls, and that's about it.

Perhaps, before we move too quickly into the service aspects of commercial refrigeration, it would be worth while to review the vapor compression cycle being used in practically all commercial refrigeration equipment.

2

Vapor Compression Cycle

The basic principle behind the vapor compression cycle is that a fluid, when changing state from a liquid to a vapor, will absorb heat in the process. Conversely, as the fluid changes from a vapor back to a liquid, it loses heat.

Refrigeration is the process of taking heat from an area where it is not wanted and moving it to an area where it is not objectionable. Combining the definition of refrigeration with the basic principle noted above, we can see that the heat taken from where it is not wanted—for example, inside a reach-in refrigerator—flows into the fluid causing the fluid to change state from a liquid to a vapor. The vapor flows through the system to a point outside the compartment, where it changes back into a liquid by giving up heat to an area where this heat is not objectionable. The liquid returns to the compartment where it is ready to once again absorb heat and repeat the cycle.

This vapor compression system is shown in Figure 2-1. The fluid, called a refrigerant, flows through the refrigerant flow control into the evaporator. The liquid in the evaporator boils off because the air passing over the evaporator is warmer than the boiling point of the refrigerant. The heat flowing into the evaporator causes the liquid to change into vapor which is drawn, through the suction line, to the compressor. The compressor increases the pressure and temperature of the refrigerant and forces it through the hot gas line, or discharge line as it is often called,

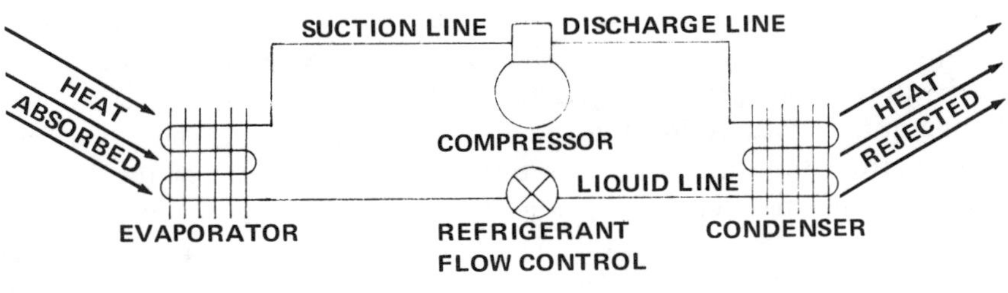

FIG. 2-1. Vapor compression refrigeration system.

to the condenser. The air passing over the condenser, being cooler than the boiling point (or condensing point) of the refrigerant, removes heat from the refrigerant and allows it to change back into a liquid. This liquid, in turn, moves through the liquid line back to the refrigerant flow control to be metered into the evaporator. The cycle is now complete.

The essential components of this cycle are:

Evaporator—the heat transfer surface that permits heat to flow into the refrigerant.

Suction Line—the path through which refrigerant flows from the evaporator to the compressor.

Compressor—the component that draws refrigerant vapor from the evaporator as fast as it is formed and compresses it, increasing its temperature and pressure.

Hot Gas or Discharge Line—the path through which the refrigerant flows from the compressor to the condenser.

Condenser—the heat transfer surface that permits heat to flow out of the refrigerant.

Liquid Line—the path through which liquid, formed in the condenser, flows to the refrigerant flow control.

Refrigerant Flow Control—the device that provides a restriction against which the compressor can pump and that meters the amount of refrigerant entering the evaporator.

Refrigerant—the fluid that absorbs heat by boiling off (or evaporating) at a low temperature and pressure and gives off heat while condensing at a higher temperature and pressure.

These are the essential components of the vapor compression system. Many refrigeration systems have only these essential components and function quite well. There are many other applications, however, where additional devices must be used to get the desired refrigeration effect. These optional devices are called accessories. While they are not essential to the basic system, they are added to increase efficiency or to protect the system or to accomplish particular control objectives.

In a few short sentences we've looked at something that is really pretty complex. For the remainder of this chapter we're going to look at the cycle quite closely, to develop a full understanding of the pressure and temperature relationships, the significance of subcooling and superheat, the determination of net refrigerating effect and the use of the Mollier diagram. This may seem like heavy stuff, but we'll approach it gradually and you'll see how it all falls together. Once you have a full understanding of how the refrigeration cycle works, you will see how this knowledge will enable you to understand the functioning of practically every other piece of commercial equipment you will ever encounter.

REFRIGERANTS

In commercial refrigeration, there are three refrigerants that you will have occasion to use. To distinguish between refrigerator and freezer applications, the terminology medium temperature and low temperature is used, with the latter referring to freezers. R-12 has found wide use in both medium and low temperature applications. R-502 is gaining increasing acceptance in low temperature applications and R-22 was used for a number of years, and is still occasionally used, in low temperature equipment.

Figure 2-2 presents a table of pressure/temperature relationships for these three refrigerants. The boiling or condensing point of a refrigerant will vary with the pressure of the refrigerant.

Looking at the chart, you will note that the boiling point of R-12,

TEMPERATURE °F.	REFRIGERANT 12	22	500	502	717
-20	0.6	10.1	3.2	15.3	3.6
-18	1.3	11.3	4.1	16.7	4.6
-16	2.0	12.5	5.0	18.1	5.6
-14	2.8	13.8	5.9	19.5	6.7
-12	3.6	15.1	6.8	21.0	7.9
-10	4.5	16.5	7.8	22.6	9.0
-8	5.4	17.9	8.8	24.2	10.3
-6	6.3	19.3	9.9	25.8	11.6
-4	7.2	20.8	11.0	27.5	12.9
-2	8.2	22.4	12.1	29.3	14.3
0	9.2	24.0	13.3	31.1	15.7
1	9.7	24.8	13.9	32.0	16.5
2	10.2	25.6	14.5	32.9	17.2
3	10.7	26.4	15.1	33.9	18.0
4	11.2	27.3	15.7	34.9	18.8
5	11.8	28.2	16.4	35.8	19.6
6	12.3	29.1	17.0	36.8	20.4
7	12.9	30.0	17.7	37.9	21.2
8	13.5	30.9	18.4	38.9	22.1
9	14.0	31.8	19.0	39.9	22.9
10	14.6	32.8	19.7	41.0	23.8
11	15.2	33.7	20.4	42.1	24.7
12	15.8	34.7	21.2	43.2	25.6
13	16.4	35.7	21.9	44.3	26.5
14	17.1	36.7	22.6	45.4	27.5
15	17.7	37.7	23.4	46.5	28.4
16	18.4	38.7	24.1	47.7	29.4
17	19.0	39.8	24.9	48.8	30.4
18	19.7	40.8	25.7	50.0	31.4
19	20.4	41.9	26.5	51.2	32.5
20	21.0	43.0	27.3	52.4	33.5
21	21.7	44.1	28.1	53.7	34.6
22	22.4	45.3	28.9	54.9	35.7
23	23.2	46.4	29.8	56.2	36.8
24	23.9	47.6	30.6	57.5	37.9
25	24.6	48.8	31.5	58.8	39.0
26	25.4	49.9	32.4	60.1	40.2
27	26.1	51.2	33.2	61.5	41.4
28	26.9	52.4	34.2	62.8	42.6
29	27.7	53.6	35.1	64.2	43.8
30	28.4	54.9	36.0	65.6	45.0
31	29.2	56.2	36.9	67.0	46.3
32	30.1	57.5	37.9	68.4	47.6
33	30.9	58.8	38.9	69.9	48.9
34	31.7	60.1	39.9	71.3	50.2
35	32.6	61.5	40.9	72.8	51.6
36	33.4	62.8	41.9	74.3	52.9
37	34.3	64.2	42.9	75.8	54.3
38	35.2	65.6	43.9	77.4	55.7
39	36.1	67.1	45.0	79.0	57.2
40	37.0	68.5	46.1	80.5	58.6
41	37.9	70.0	47.1	82.1	60.1
42	38.8	71.4	48.2	83.8	61.6
43	39.8	73.0	49.4	85.4	63.1
44	40.7	74.5	50.5	87.0	64.7
45	41.7	76.0	51.6	88.7	66.3
46	42.6	77.6	52.8	90.4	67.9
47	43.6	79.2	54.0	92.1	69.5
48	44.6	80.8	55.1	93.9	71.1
49	45.7	82.4	56.3	95.6	72.8
50	46.7	84.0	57.6	97.4	74.5
55	52.0	92.6	63.9	106.6	83.4
60	57.7	101.6	70.6	116.4	92.9
65	63.8	111.2	77.8	126.7	103.1
70	70.2	121.4	85.4	137.6	114.1
75	77.0	132.2	93.5	149.1	125.8
80	84.2	143.6	102.0	161.2	138.3
85	91.8	155.7	111.0	174.0	151.7
90	99.8	168.4	120.6	187.4	165.9
95	108.2	181.8	130.6	201.4	181.1
100	117.2	195.9	141.2	216.2	197.2
105	126.6	210.8	152.4	231.7	214.2
110	136.4	226.4	164.1	247.9	232.3
115	146.8	242.7	176.5	264.9	251.5
120	157.6	259.9	189.4	282.7	271.7
125	169.1	277.9	203.0	301.4	293.1
130	181.0	296.8	217.2	320.8	—
135	193.5	316.6	232.1	341.2	—
140	206.6	337.2	247.7	362.6	—
145	220.3	358.9	264.0	385.0	—
150	234.6	381.5	281.1	408.4	—
155	249.5	405.1	298.9	432.9	—

FIG. 2-2. Saturated pressure/temperature table for commonly used refrigerants. *Courtesy Sporlan Valve Co.*

under a pressure of 25 psig, is about 26°F, while this same refrigerant, at a pressure of 125 psig, has a boiling point of about 104°F. Refrigerant 502 has boiling points of about —7°F and 65°F at the same pressures, while R-22 has boiling points of 2°F and 72°F. Accuracy within a degree or two is all that we need in any practical problem.

These pressure/temperature relationships are presented here in tabular form and many companies print these tables on handy cards that the serviceman can slip into his pocket. The proper use of the card can provide you with a valuable diagnostic tool. The information on the card is only part of the information contained on a Mollier diagram. We're going to look at the Mollier Diagram shortly, but first let's see what else we have to know about refrigerants.

Probably as important as any other property, from the practical service standpoint, is the interaction of refrigerant and water. Water in a refrigeration system has two consequences. Water combining with refrigerant can form a mild acid which, over a period of time, can decrease the lubricating effect of the oil in the system, causing excessive wear and premature failure. Just how premature the failure will be depends upon

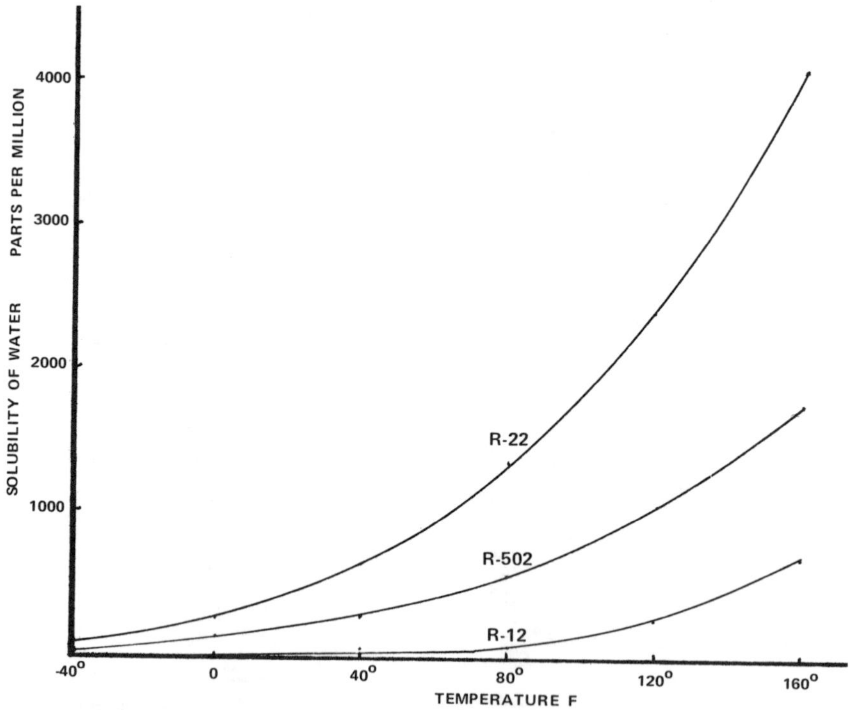

FIG. 2-3. Solubility of water in liquid refrigerant.

many factors such as operating temperatures, *on* time, system design, etc.

The acid can also attack the insulation on the motor windings of full hermetic compressors. In time, the insulation can break down, causing short circuits and possibly compressor burnout. A final effect, though not too common, is that it attacks the metal parts of the system and can cause mechanical failure.

The second consequence of too much water in the system is that undissolved or free water will freeze at the refrigerant flow control and cause a blockage. How much is too much? Well, R-12 at 78°F can hold 93 parts of water per million parts of refrigerant, R-502 can hold 560 ppm and R-22 can hold 1300 ppm, Figure 2-3. What this means is that an R-12 system with 10 pounds of refrigerant can tolerate about 1/100 oz. of water, a similar sized R-502 system can hold 9/100 oz. of water and a R-22 system 20/100 oz. before *free* water is present and a freeze-up will occur. It is essential, therefore, that we handle refrigerant in a way that minimizes the introduction of water vapor.

VAPOR COMPRESSION CYCLE REVISITED

Let's take another look at the vapor compression cycle and start to build on what we know thus far.

In the evaporator, the liquid refrigerant is absorbing heat from the surroundings and changing to a vapor. The temperature at which this boiling occurs depends on the pressure in the evaporator and the refrigerant being used. We'll assume R-12 is being used and we've connected a gage that shows a pressure of 21 psig in the low side of the system. Referring to the pressure/temperature table, we can see that the refrigerant is boiling off at 20°F. As the refrigerant flows through the evap-

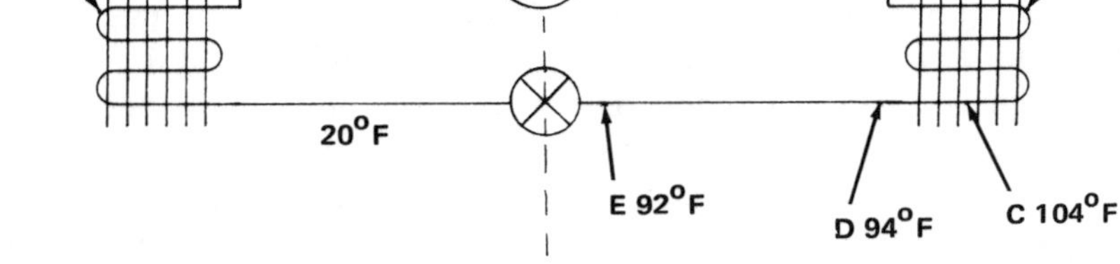

FIG. 2-4. Typical temperatures and pressures in a system operating with R-12 refrigerant.

orator, more and more liquid changes to vapor at a constant 20°F. At point X, in Figure 2-4, all of the liquid refrigerant has boiled off into a vapor at 20°F. As the vapor flows through the remaining part of the evaporator, which is approximately 5% of the total evaporator surface, it picks up additional heat and increases in temperature to 30°F, at point Y. Notice that wherever liquid is boiling, changing to vapor, the temperature is a constant 20°F, the boiling temperature, as determined by the pressure of 21 psig. As soon as the last bit of liquid has changed to vapor, at point X, any additional heat absorbed by the refrigerant produces a change in temperature. At point Y, the actual temperature of the refrigerant was measured and found to be 30°F. This is 10°F above the boiling point. The refrigerant at point Y is said to be superheated 10°F. Superheat then is the number of degrees a refrigerant is heated above its boiling point. Since the refrigerant at point Y is heated above its boiling point (superheated) we can be assured that all of the liquid is in the vapor state.

The compressor is a vapor pump. It takes in a low temperature vapor and discharges a high temperature vapor. If liquid refrigerant gets into the compressor, liquid being incompressible, there will be a sudden stoppage causing bent and broken parts and, of course, a cessation of the refrigeration process.

The compressor has drawn in vapor from the suction line at 33°F (in our example, the vapor flowing through the suction line picked up some more heat increasing its temperature) and 21 psig and compressed this vapor to 125 psig at a temperature of 122°F measured in the hot gas line. Referring to the vapor pressure table, Figure 2-2, we see that the boiling point (or condensing point) at this pressure is 104°F. Since the actual temperature of the refrigerant measured at the hot gas line is 122°F, the gas is superheated 18°F. In order for condensation to start, the refrigerant must be cooled (desuperheated) to the condensing temperature of 104°F. This desuperheating occurs in the hot gas line, to some extent, but mostly in about the first 5% of the condenser. At point B of Figure 2-4, the refrigerant has cooled to 104°F and condensing starts. As the refrigerant flows toward point C, more and more of the vapor changes to liquid until, at point C, all of the refrigerant has become liquid, still at 104°F. The amount of condenser surface used to condense the refrigerant is about 90%. The refrigerant now flows through the remaining 5% of the condenser as a liquid. Since there is no more vapor that can give up heat and change to liquid, any heat removed from the liquid refrigerant results in a drop in temperature. At point D, the temperature has dropped to 94°F, 10° under the condensing temperature of 104°F. This refrigerant is said to be *subcooled* 10°F. Subcooling then is the number of degrees *liquid* refrigerant has been cooled *below* its condensing temperature (boiling temperature).

You might have noticed that we've been using boiling point and condensing point interchangeably because they represent expressions for the same temperature. From this point on we will use the expression *saturation temperature* to describe this point. If a refrigerant is at its saturation temperature, a portion of it is in the liquid state and a portion of it is in the vapor state. If heat is added to the refrigerant, some liquid will change to vapor; if heat is taken away, some vapor will change to liquid and the temperature will remain *unchanged.* If the refrigerant liquid/vapor mixture has absorbed enough heat to change it completely to vapor, any addition of heat will result in an increase in temperature and the refrigerant is said to be *superheated.* (Superheating is occurring between X and Y in evaporator.) If the liquid/vapor mixture has given up enough heat to change it completely to a liquid, any further removal of heat will result in a decrease in temperature and the refrigerant is said to be *subcooled.* (Subcooling is occurring between C and D in the condenser.)

The subcooled refrigerant leaves the condenser and flows through the liquid line, as a liquid, to the refrigerant flow control. Further subcooling occurs in the liquid line and the refrigerant enters the flow control at E at a temperature of 92°F. The refrigerant flow control is actually a restriction in the line through which the refrigerant must pass. The flow control is designed to pass the proper amount of *liquid refrigerant* for the system in which it is installed.

If the flow control has been designed for a liquid flow of 10 pounds of R-12 per minute, you can see how the presence of vapor bubbles in the line could very well reduce that flow, Figure 2-5. The bubbles in a liquid stream are given the name of *flash gas.* The presence of flash gas in the liquid line cuts down the flow of refrigerant and therefore reduces the refrigeration effect of the system. From our definition of subcooling, vapor cannot exist in a subcooled refrigerant. If we design the system to assure that only subcooled refrigerant is in the liquid line, we can be confident that we will have 100% liquid up to the refrigerant flow control.

As the refrigerant passes through the refrigerant flow control, it passes from an area of high pressure (125 psig) to low pressure (21 psig). Refrigerant at a temperature of 92°F (Figure 2-4, point E) is moving into an area where the saturation temperature is 20°F. *Liquid* cannot exist at a temperature above its saturation point. A portion of the liquid immediately boils off into a vapor. This boiling process results in cooling of the refrigerant. That is, a portion of the warm liquid rapidly boils off, or flashes, absorbing heat from the rest of the refrigerant. A typical

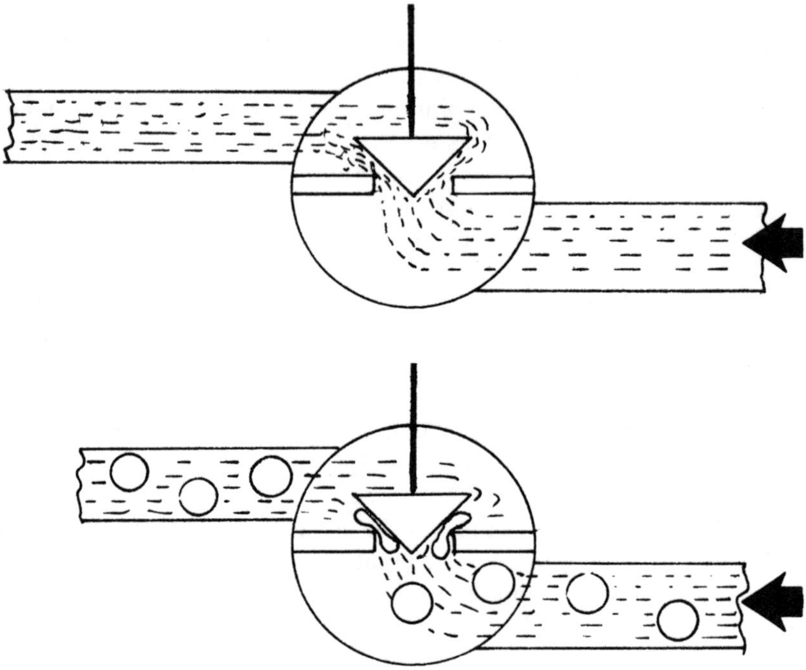

FIG. 2-5. Subcooled liquid, above, assures maximum refrigerant flow and maximum cooling. Flash gas, below, displaces liquid in the line and reduce the amount of refrigerant flowing to the evaporator.

example of this process shows about 2 pounds of refrigerant flashing off out of every 10 pounds passing through the flow control. Immediately after leaving the valve, the refrigerant is about 20% vapor and 80% liquid. This 20% flash gas does no useful cooling; only the 80% remaining liquid is available to absorb heat and change to a vapor.

Flash gas leaving the refrigerant flow control is unavoidable in the typical refrigeration cycle. Flash gas in the liquid line approaching the refrigerant flow control must be avoided, and can be avoided, in order for the system to produce the designed cooling effect.

Now the vapor/liquid mixture moves into the evaporator and the cycle is complete.

3

TXV Systems

The TXV (thermostatic expansion valve) is the most widely used refrigerant flow control in commercial refrigeration applications. TXV's come in all sizes and shapes, Figure 3-1, but all have the same function: to maintain a constant superheat in the suction line to the compressor.

Figure 3-2 shows the circuit layout of the TXV system. You will note that it resembles Figure 2-1 with the exception that the TXV is specified as the refrigerant flow control and that a receiver, an accessory, has also been added.

The receiver, Figure 3-3, has several functions in the system: (1) it stores excess refrigerant, permitting us to add somewhat more than the

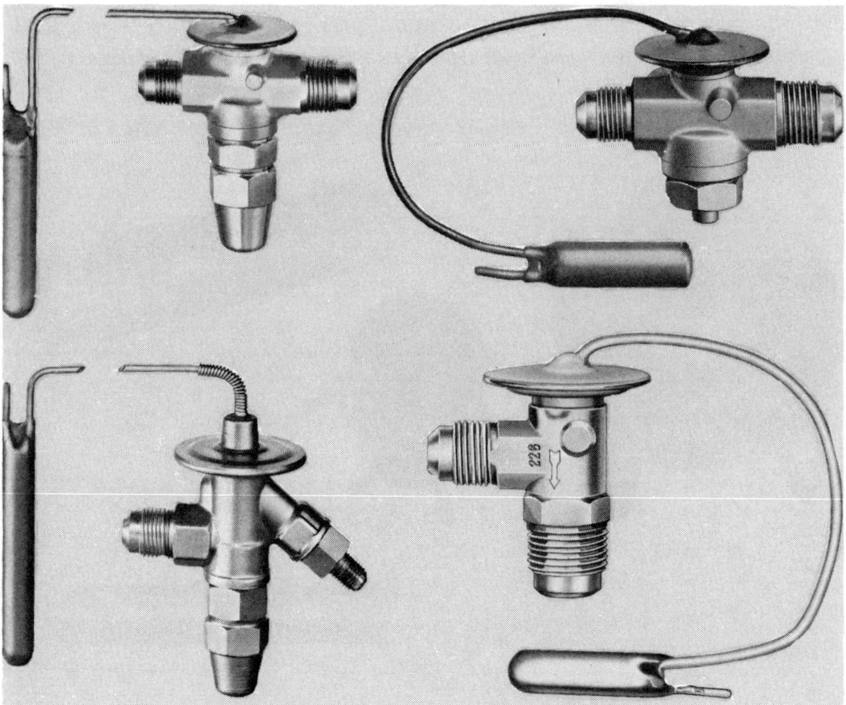

FIG. 3-1. Three of these expansion valves are field adjustable. The valve at the bottom right is not. *Courtesy Controls Div., Singer Co.*

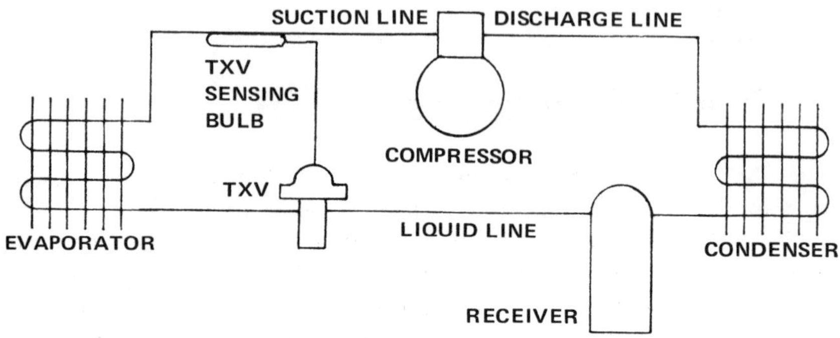

FIG. 3-2. Schematic of a TXV system.

absolute minimum refrigerant charge. If there is a small leak in the system (several ounces a year), the system can run several years before repairs will be needed. (2) It provides a place to store refrigerant when pumping down the system prior to maintenance. (3) It provides a seal at the liquid line, Figure 3-4, to assure that only liquid gets to the TXV.

FIG. 3-3. Receivers are available in several configurations. *Courtesy Lehigh, Inc.*

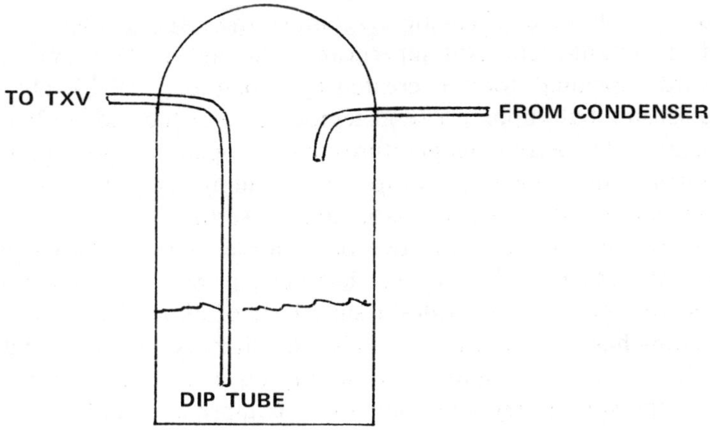

FIG. 3-4. Refrigerant in the receiver must cover the dip tube. Otherwise, bubbles will appear in the sight glass.

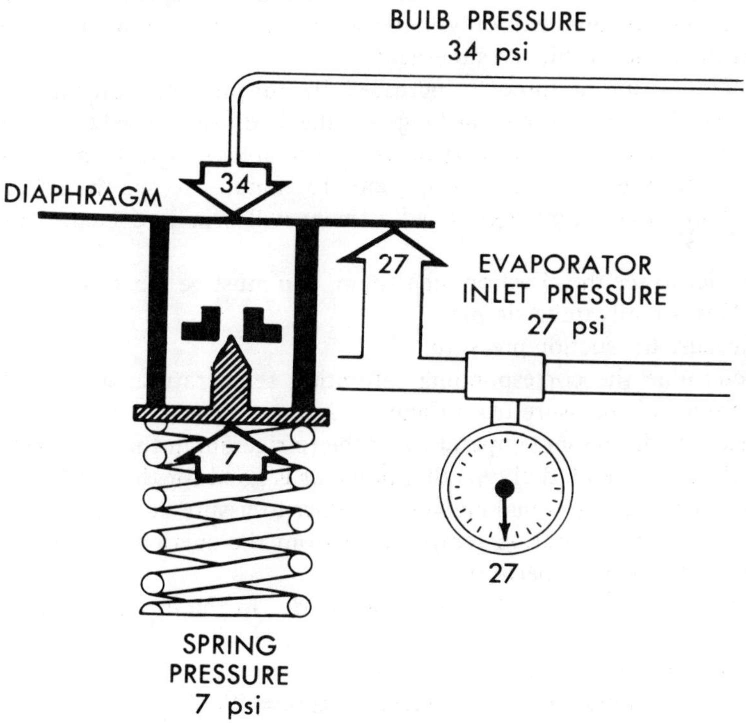

FIG. 3-5. Forces acting to operate the TXV. *Courtesy Sporlan Valve Co.*

SUPERHEAT AND THE TXV

Figure 3-5 shows a schematic diagram of the forces acting within the TXV to maintain a constant superheat at the end of the evaporator. A downward (opening) force is created by expansion of the vapor in the remote sensing bulb upon a rise in temperature at the suction line close to the evaporator. Rising temperatures tend to open the valve, permitting more refrigerant to enter the evaporator. Falling temperatures indicate that adequate refrigerant is available and close the TXV.

Counteracting this force are two other forces. One is the evaporator pressure, the other is a spring that has been preset at the factory. The spring setting generally provides a superheat of about 10° unless some other setting has been specified. This is a satisfactory superheat and there should be no reason to adjust this setting on a new valve. With time, however, the spring may lose some of its compression and may require adjustment as it gets older.

Looking again at Figure 3-5, you can see that the spring tends to close the valve, restricting refrigerant flow. If we increase the compression on the spring, a large downward force is required to keep the valve open. This downward force will only increase with an increase in suction line temperature, hence higher superheat.

Generally, spring force is increased by turning the adjusting stem clockwise. If there are no markings on the body of the valve, you can make this assumption. A good procedure to follow is to turn the stem about ½ turn at a time and wait about 15 minutes to see the effect on the system. A counterclockwise adjustment will usually decrease superheat.

In order to see the effect on the system, you must be monitoring superheat. This is really quite simple.

1. Measure the suction pressure.
2. Determine the corresponding saturation temperature of the refrigerant at this pressure from Figure 2-2.
3. Measure the actual temperature of the refrigerant at the TXV sensing bulb. (A bimetal, dial type thermometer is good for this.) Make sure the thermometer is making good contact to assure accuracy.
4. Subtract the saturation temperature from the suction line temperature. This is the superheat.

This takes longer to tell than it does to do. In the example shown in Figure 3-6:

1. Suction pressure = 27 psig
2. The corresponding saturation temperature = 28°F
3. The temperature at the sensing bulb = 37°F
4. Superheat = 37° − 28° = 9°

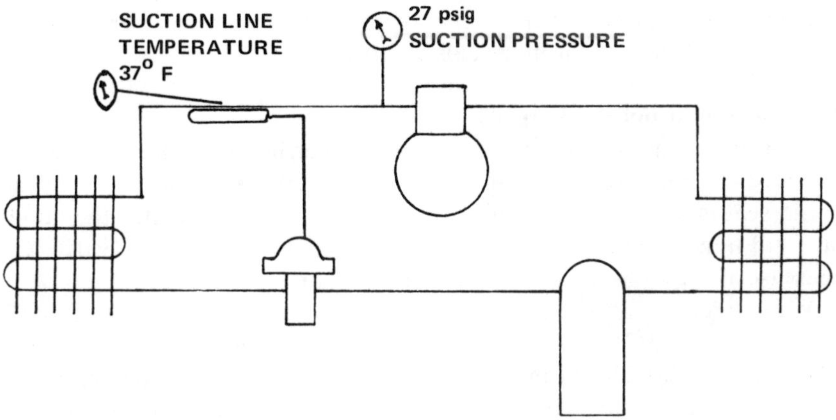

FIG. 3-6. Measuring superheat.

Superheat monitoring is a valuable service technique because it gives you an indication of how efficiently the evaporator is being used. If you get high superheat readings, on the order of 20° or more, it is an indication that not enough refrigerant is getting into the evaporator. This could be caused by an improperly set TXV, but before attempting to adjust the valve, look at these other areas:

1. Is the unit properly charged?
2. Is there a restriction in the line?
   (a) Kinked liquid lines?
   (b) Clogged filter drier?
   (c) Freeze-up?
3. Is the condenser coil clean, permitting maximum compressor efficiency?

The source of the problem could be, and usually is, in one of these areas. Only after checking these out should you attempt to adjust the valve; not because adjusting the valve is hard, but because the valve is the least likely source of the problem and adjusting it properly is a time-consuming procedure.

On the other hand, if you get low superheat readings, the problem is almost invariably associated with an improperly set valve or a sensing bulb not securely attached to the suction line. Once it is determined that the bulb is securely attached and making good metal-to-metal contact, adjustment of the superheat setting should be attempted, monitoring superheat throughout the procedure.

If continued efforts at adjustment do not correct the situation, the valve should be replaced.

Since high superheat can be the symptom of several rather common problems, let's look a bit more closely at these:

1. Is the unit properly charged?

A properly charged unit will have enough refrigerant to fill the receiver to a level over the dip tube, Figure 3-4. If the level is below this tube, the liquid seal is lost and a mixture of liquid and vapor will enter the liquid line and flow to the TXV. The flow of refrigerant through the valve will decrease if vapor is present, resulting in high superheat.

We can tell if vapor is mixed with the liquid if we have a sight glass in the line, Figure 3-7. Bubbles of vapor in the sight glass usually mean that we've lost the liquid seal in the receiver and the unit is undercharged.

What if there is no sight glass in the system? If you plan to service this equipment on a regular basis, install a sight glass. It makes life so much simpler.

There are two other techniques to use. First, check the high side pressure. The saturation temperature on the high side of the system is generally 20° to 30°F higher than the entering air temperature. If the entering air temperature is 90°F, add 30°, for a high side saturation temperature of 120°F. Referring to Figure 2-2, the high side pressure corresponding to 120°F is about 157 psig. If your measured high side pressure is more than 15 to 20 psig lower than this pressure, the system is most likely undercharged. Not all systems are the same and this tech-

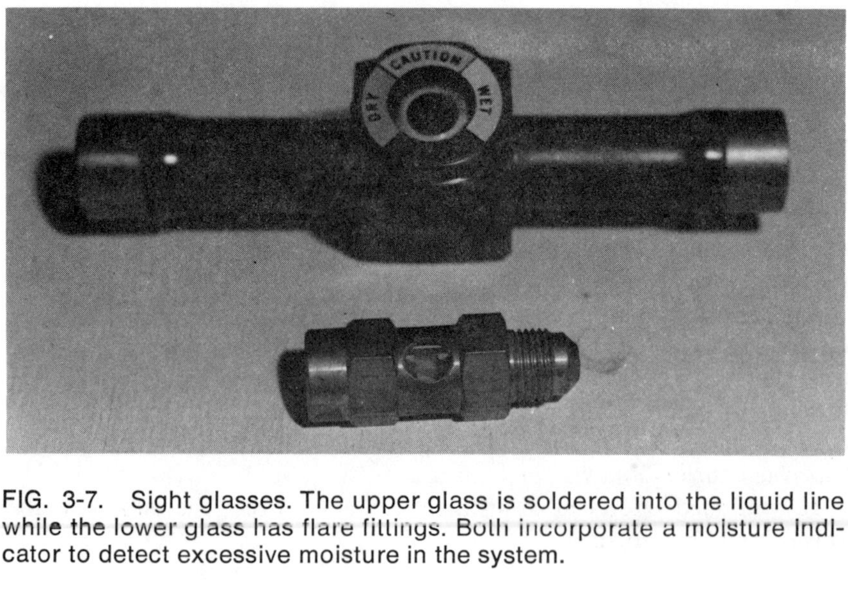

FIG. 3-7. Sight glasses. The upper glass is soldered into the liquid line while the lower glass has flare fittings. Both incorporate a moisture indicator to detect excessive moisture in the system.

nique is just an approximation, but it will give you some idea as to the adequacy of the charge.

The second technique is to check the subcooling at the TXV inlet. The liquid line should show 5° to 10° subcooling at the TXV to guarantee 100% liquid. You can determine subcooling as follows:

(1) Measure the high side pressure.
(2) Determine the saturation temperature from Figure 2-2.
(3) Measure the actual temperature of the liquid line at the inlet of TXV.
(4) The number of degrees that the actual temperature is below the saturation temperature is the subcooling. Remember, this technique is accurate *only* if there are no abnormal restrictions in the liquid line.

2. Is there a restriction in the liquid line?

A kinked liquid line can restrict refrigerant flow. With a restriction, there may be enough pressure drop to cause some liquid to flash off into vapor.

Suppose the refrigerant liquid leaving the receiver was at a temperature of 120°F and at a pressure of 157 psig. Checking our pressure/temperature chart, we see that this is a saturated condition. If a kink in the liquid line creates a 10 psig pressure drop, the refrigerant pressure will be 147 psig after the kink. The boiling point of R-12 at 147 psig is 115°F. What happens when 120°F refrigerant enters an area where its boiling point is 115°F? *Flashing will occur* until enough liquid boils off to a vapor to

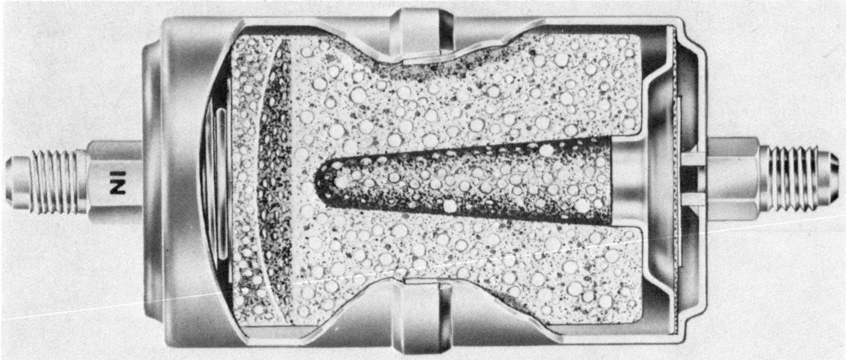

FIG. 3-8. Cutaway of a filter-drier shows the dessicant used to remove moisture from the system. Filter-driers are position sensitive and must be installed as indicated by the manufacturer. In this case, denoted by IN. *Courtesy Henry Valve Co.*

bring the temperature of the refrigerant down to 115°F. This flash gas in the liquid line cuts down on the capacity of the TXV and we have a high superheat condition. Is the TXV at fault? No! Is there a shortage of refrigerant charge? No!

An interesting thing to note is that, if the restriction occurs upstream (on the condenser side) of the sight glass, bubbles will appear in the glass. Remember: bubbles in the sight glass *do not always* mean the system needs refrigerant.

Once we've established that the liquid line is not kinked, we should look at one other obvious source of restriction in the liquid line, namely, the filter-drier, Figure 3-8. A filter-drier saturated with moisture will not necessarily create a restriction. However, as dirt is filtered out and builds up in the filter-drier, the restriction or pressure drop increases gradually until it eventually becomes significant and large amounts of flash gas form, Figure 3-9. Most filter-driers are sized to offer a restriction or pressure drop of about 2 to 3 psi. As this pressure drop increases to 10 to 12 psi, because of dirt buildup, flashing will occur. On severe cases, and you'll run into many severe cases, the refrigerant temperature will drop enough for you to feel it with your bare hand, and you may even see that the drier is sweating. If the liquid line up to the drier is warm

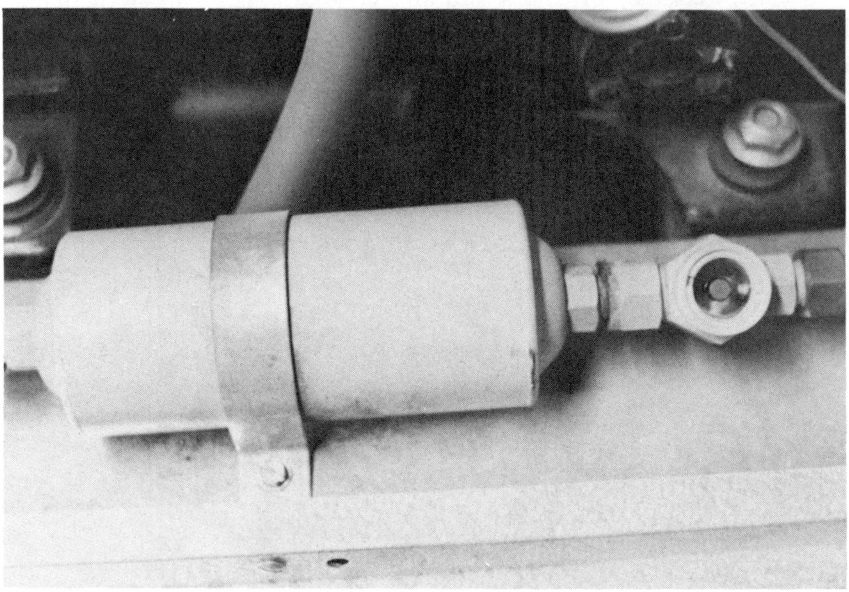

FIG. 3-9. The sight glass must always be downstream from the filter-drier if it is to indicate a restriction through the drier. In this case, flow is from left to right.

and after the drier is cold, that's a sure sign that the drier is creating an excessive restriction and should be replaced.

Another restriction in the system is created by excess moisture accumulating as free water at the expansion device and freezing up. In refrigerators and freezers, the refrigerant temperature will normally be below 32°F as it leaves the TXV, therefore free water will freeze at the valve orifice. Once this happens, refrigerant flow is either fully or partially restricted resulting in, you guessed it, high superheat.

Where does this water come from? Usually systems shipped from the plant have been dehydrated to the *n*th degree and have practically no free water. So confident are many manufacturers about their dehydrating techniques that they don't install a filter-drier on the unit. This is not too good for the serviceman. No matter how carefully you purge your gage lines; no matter how careful you are in attaching the gage manifold; no matter how diligent you are about adding clean, dry refrigerant, the sad fact is that everytime you open a refrigeration system you introduce moisture, in infinitesimal amounts perhaps, but some moisture in any event. Over the years, as more and more servicemen attach their gages, the moisture level builds up until the ability of the refrigerant to absorb it is exceeded. Free water is now present and freeze-up will occur.

Can a freeze-up occur in a system never touched by human hand? Yes. Full hermetic compressors, and semi-hermetics for that matter, have motor windings exposed to the refrigerant flow. The windings are covered with insulation that has water in it. Under normal operating conditions the water stays in the insulation. What happens if the compressor motor runs hot due to high superheat (which of course means poor suction gas cooling) or due to high head pressures (because of a dirty condenser or a warm condenser location)? Why, of course, the insulation gets hot and dries out. Drying out means that the water is driven out of the winding insulation into the refrigerant surrounding the windings. Result? Freeze-up. It may take 5, 10 or even 20 years, depending on conditions surrounding the unit, but it can happen.

You might wonder how you know you have a freeze-up and what you do about it. Once you've established that you have no restrictions and enough refrigerant, a high superheat reading can only mean a bad TXV or a freeze-up. Treat the TXV as though it was frozen. If that doesn't work, change the valve.

A freeze-up is usually accompanied by low suction pressure. When the compressor is shut down, the pressure will remain low and might gradually inch upward. By wrapping a warm wet cloth around the TXV, you should be able to melt the blockage.

Never apply a direct flame to the body or diaphragm assembly of a

TXV. You may not see it, but chances are you'll do severe damage to the internal workings of the valve.

When heat on the body of the valve causes a rapid pressure climb, you have a freeze-up. After melting the freeze-up, start the system. Within a matter of minutes pressures will look good, then suddenly the suction pressure will pull down into a vacuum, indicating the freeze-up has recurred.

To remove the moisture:

(a) Replace the filter-drier.

(b) If there is no filter-drier, install one.

(c) Permit the system to run until you are convinced a freeze-up will not recur. Usually, recurrence will take less than 5 minutes. If your system runs 15 to 20 minutes without freeze-up again, chances are you've got it beat.

There are dehydrating fluids, or antifreezes, on the market for refrigeration systems. You can have good experience with them if you follow the directions on the bottle. Generally 1/8 oz. of antifreeze per pound of refrigerant is all that is required (that's about 1 part antifreeze per 100 parts refrigerant). *You cannot use this stuff on systems with a filter-drier.* The desiccant in the drier will absorb the antifreeze and, believe it or not, release the water, leaving you worse off than you were in the first place. Some manufacturers claim they will void their warranty if they find traces of antifreeze in returned compressors. Be smart, use this good stuff on out-of-warranty compressors and go to the extra effort of installing a drier on your in-warranty jobs.

If, after treating the system for a freeze-up, you find that the valve is in fact no good, obviously, you have to replace the valve. While you're at it, change the drier too. Do the job right.

3. Is the condenser coil clean, for maximum condenser efficiency?

An evaporator and condensing unit are matched by the manufacturer to do a specific job. The ability of the equipment to do this job depends, to a great degree, upon its cleanliness. A dirty condenser coil loses its ability to reject heat. The only way the heat, that was absorbed by the refrigerant in the evaporator, can be rejected to the outside air is by having the compressor create a higher temperature and pressure in the dirty condenser. The compressor will, in fact, do this without any urging from you. However, in the process of working harder, the compressor will lose efficiency and pump less refrigerant.

There is a way we can predict what will happen to a refrigeration cycle under various conditions. The Mollier diagam has proved to be a useful tool for really good service engineers. We will take a look at the Mollier diagram a little later and see how it can help us recognize conditions that may be hazardous to the refrigeration system.

Meanwhile, you can take it on faith that several things will happen to our system:

(a) The discharge temperature will increase, possibly causing the oil to break down.

(b) Higher condenser pressures mean less refrigerant being pumped, therefore lower capacity.

(c) Higher suction line temperature and a reduction in compressor cooling.

(d) Shorter compressor life.

EVAPORATOR PROBLEMS

The evaporator is the surface that absorbs the heat from the refrigerated space. The efficiency of this surface is directly related to the velocity at which air passes over it.

Most commercial reach-in refrigerators use a blower coil, that is, have a fan blowing air across the surface. Some, however, use a gravity coil, utilizing natural convection to create air movement in the box.

FIG. 3-10. A typical evaporator blower motor. It combines low cost with long, service-free life. *Courtesy Motors & Armatures, Inc.*

If air movement is less than required, several things will happen:

1. The cooling capacity of the system will be reduced.
2. Ice may form on the evaporator, resulting in still less cooling capacity.
3. Liquid refrigerant may flood back to the compressor causing serious damage.

There are two major causes of reduced air flow over blower coils: dirt and fan motor failure.

Depending on what is stored in the refrigerator, a substantial buildup of dirt can occur over a period of years, with a gradual erosion of cooling capacity until, at some point, frost starts to accumulate and cooling performance is severely affected.

The little motors used in reach-in refrigerators, Figure 3-10, are really amazingly durable. They will run for years and years without a drop of oil. In fact, many don't even have provision for lubrication. At some point, however, the bearings will wear and seize causing zero air flow. Usually this is a total failure and quite easy to see. Since the evaporator fan motor in almost all reach-in refrigerators runs 100% of the time, a simple visual check is usually all that's required to see if the evaporator fan is the source of a problem.

A gravity coil, with no fan, will rarely get dirty enough to affect capacity and, of course, there is no motor to malfunction. Assuming all the controls are functioning properly, the major cause of problems in a box with this type of coil is improper packing. Since air motion is slow, any major obstruction will impede its flow. Completely filling the box, permitting no free circulation of air, or shelf paper or foil used to keep the wire shelves clean, can easily inhibit the flow of air to the point of decreasing capacity and eventually causing excessive ice buildup.

CONTROLS

The control with which we are mainly concerned in reach-in refrigerators is the one that maintains box temperature within required limits. Both thermostats and pressurestats are used for this purpose, but the thermostat is used with cap tube systems and the pressurestat with TXV systems, Figure 3-11.

The pressure control takes advantage of the fact that there is a definite, reproducible relationship between the temperature of the refrigerant in the evaporator and the temperature of the air passing over the evaporator. Temperature differences on the order of 12° to 15° are typical. If, for example, the air temperature of a particular box is 35°, then the refrigerant temperature will be, let's say, 22°F. For this particular box, every time the air temperature is 35°F, the refrigerant temperature will be 22°F. Another refrigerator of the same model may have a slightly

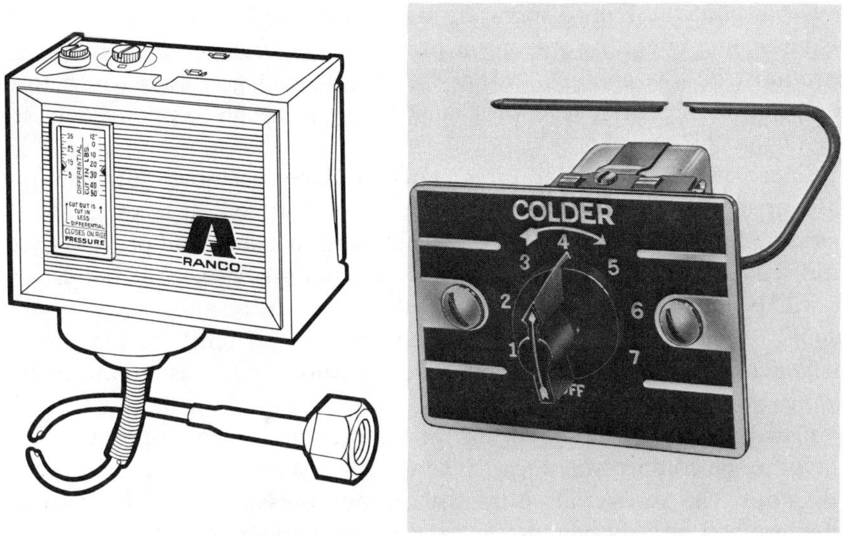

FIG. 3-11. Box temperature is maintained by a pressurestat (left) with TXV systems and by a thermostat (right) with cap tube systems. *Courtesy Ranco Controls Div.*

different relationship, but once that relationship exists, it will be reproducible.

It's easy to relate the box temperature to refrigerant pressure using the pressure/temperature chart, Figure 2-2. In the example we've been studying, the 22°F refrigerant (assume R-12) will be at a pressure of 22.4 psig. Therefore, if we have the pressurestat set to cut out at 22.4 psig, every time the box temperature is reduced to 35°F the pressure control will open, shutting off the compressor. By reducing the cut-out pressure to a lower value, a colder temperature can be reached; by increasing the cut-out setting, a warmer temperature can be maintained.

While pressure controls are provided with a scale and graduations in psig, the indicators are not always reliable. The only sure way to set the cut-out on a pressure control is to use a pressure gage. Incidentally, since the box temperature/refrigerant temperature relationship is not specified, and in fact varies between boxes, a thermometer should be used to measure box temperature in conjunction with the pressure gage. When you have set the control to cut out at the desired box temperature, note this cut-out pressure somewhere on the unit, so that on any subsequent service calls you can check the control for proper functioning.

By setting the cut-out pressure, we have established the lower temperature limit. Next, we'll set the cut-in pressure and the control will, from now on, hold the box temperature within the two limits. There are two requirements in setting the cut-in pressure/temperature: first, the pressure must be low enough, so that the box doesn't get too warm before the compressor starts; second, it must be high enough to permit defrosting of the coil.

Since refrigerant temperatures can very likely get down into the low 20's or teens, we can expect ice to form on the coil surfaces. If ice is permitted to build up, it will begin to act as an insulator and prevent good heat transfer between the air and the refrigerant. Also, as it builds up, it will begin to obstruct air flow, still further decreasing capacity. We would like our control to keep the compressor off until the ice has all melted from the evaporator coil. The pressure control is a natural for doing exactly that.

Figure 3-12 shows a plot of refrigerant pressure in the evaporator from the time the compressor stops to the time it restarts. When the compressor stops, the refrigerant is at the cut-out pressure (A). Refrigerant pressure will quickly increase a few psi until it reaches the pressure corresponding to the coil temperature (B).

In this example, the coil temperature, and the ice on the surface, is 26°F. The pressure has increased quickly from 22.4 psig (refrigerant temperature is 22°F) to 25.3 psig (coil surface temperature is 26°F). Since 35° air is passing over the surface of the coil, you would expect the 26° coil and the 26° ice to warm and, in fact, they do. As the coil warms up, the refrigerant in the coil also gets warmer. The refrigerant in the coil will *not* get warmer than the surface of the coil. When the coil has increased in temperature to 32°F, the pressure will be 30.0 psig (C). The ice will now start to melt. The plot in Figure 3-12 shows a plateau (C-D) at this point. The pressure will not begin to increase again until all the ice has melted. Once the last ice has melted, the pressure increases rather rapidly (E). It will increase as the coil warms up toward the box air temperature. While this melting has been occurring, heat has been leaking into the refrigerator through the walls, because of door opening and closing, and perhaps through added product being placed in the box. The air temperature will have gone up one or two or more degrees, depending on the amount of heat leakage. If the box is 39°F, the evaporator pressure will rise toward 36 psig. If we set the cut-in pressure over 30 psig, we will fulfill the defrosting requirement. If we set it at less than 36 psig, we will prevent the box temperature from going over 39°F. In the absence of any manufacturer's instructions to the contrary, a good cut-in pressure will be about 33 to 34 psig.

While the cut-out pressure may vary from box to box and model to model, depending on its design, the cut-in pressure is primarily estab-

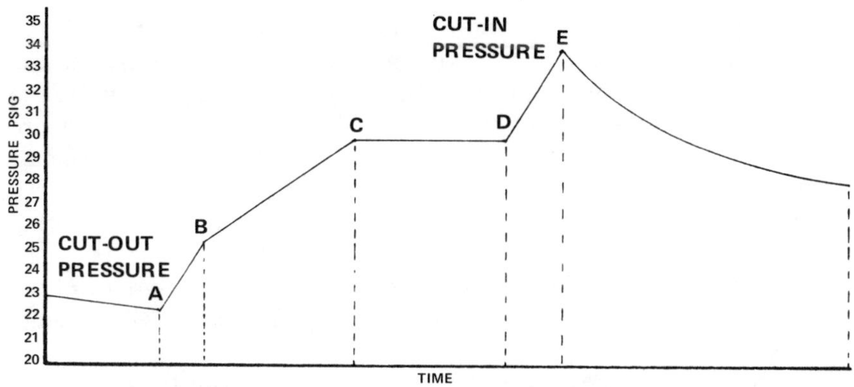

FIG. 3-12. Pressurestat-controlled, self-defrosting cycle.

lished by the defrosting requirement. In fact, the cut-in pressure of 33 to 34 psig can be used not only in reach-in refrigerators, but in walk-ins as well.

Just as a reminder: set the cut-in pressure using a pressure gage. Don't rely on the scale on the control.

Figure 3-13 shows the scale found on the face of a typical pressure control. In our example, we had a cut-in of 34 psig and a cut-out of 22.4 psig. Differential is the difference between the cut-in and the cut-out. In our case, the differential is 34 minus 22.4 or 11.6 psi. Therefore, the control will be set for a cut-in of 34 and a differential of 11.6. Setting the control this way will get you close. The final adjustment is made, as previously mentioned, with a pressure gage so that you can actually see the cut-in and the cut-out pressures on the gage.

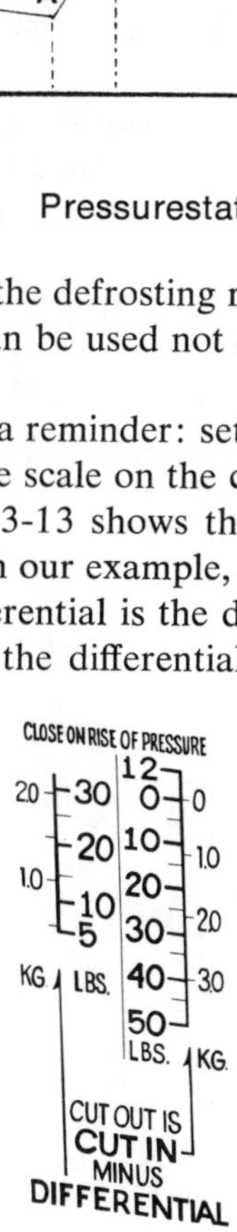

FIG. 3-13. Close-up of a low-pressure control face that is scaled in both psig and kilograms/square centimeter. Cut-in is to the right, differential at the left. *Courtesy Ranco Controls Div.*

Instead of cut-in and differential settings, other controls may use cut-in and cut-out settings. This should pose no problem, in fact it is a bit more straightforward. In either case, remember to use your gages to arrive at the final setting.

While many systems utilize a pressure control to maintain temperature, many others use thermostats. Typically, the sensing bulb of the thermostat is attached to the tubing of the evaporator coil. In this way, it is actually measuring the coil surface temperature rather than the air temperature. This being the case, we have a great deal of similarity in the operation of the pressure control and the thermostat. Since we've seen the relationship between air temperature and refrigerant temperature, you can see that measuring the refrigerant temperature can be just as reliable a guide as measuring pressures.

Some thermostat sensing bulbs may be mounted on the surface of the coil rather than the tubing itself. Again, the performance will be quite similar.

In most cases where thermostats are used, the cut-in temperature is fixed and the cut-out temperature can be adjusted to produce a higher or lower temperature in the box. The cut-in is fixed sufficiently high, to produce the self-defrosting action previously described.

So far we've looked at reach-in refrigerators utilizing thermostatic expansion valves as the refrigerant flow control. This is quite an important application because probably more TXV systems, of all sizes, are in existence in the commercial field than any other type. The *any other type* to which we refer is the cap tube. There are enough differences between cap tube and TXV systems to warrant another chapter.

4

Capillary Tube Systems

There are a great many reach-in refrigerators designed around the capillary tube refrigerant flow control, Figure 4-1. The primary reason for using a capillary tube, rather than a TXV, is cost. If a TXV costs $15 for a particular refrigerator, a capillary tube will cost about $2. On the face of it, this is not an insignificant savings, but this is only part of the story.

Cap tube systems do not require the high starting torque compressors needed in TXV systems, that's a few dollars. Low torque compressors don't use start capacitors, that's a few dollars. Low torque compressors generally use current relays instead of potential relays, that's a few dollars. Cap tube systems don't use the receivers needed by TXV systems, and that's a few dollars. If we add up all those little cost savings, they translate into a substantial price advantage over TXV systems. Not only that, but they work pretty well, too.

If you refer back to Figure 3-5, you'll see the forces that make the

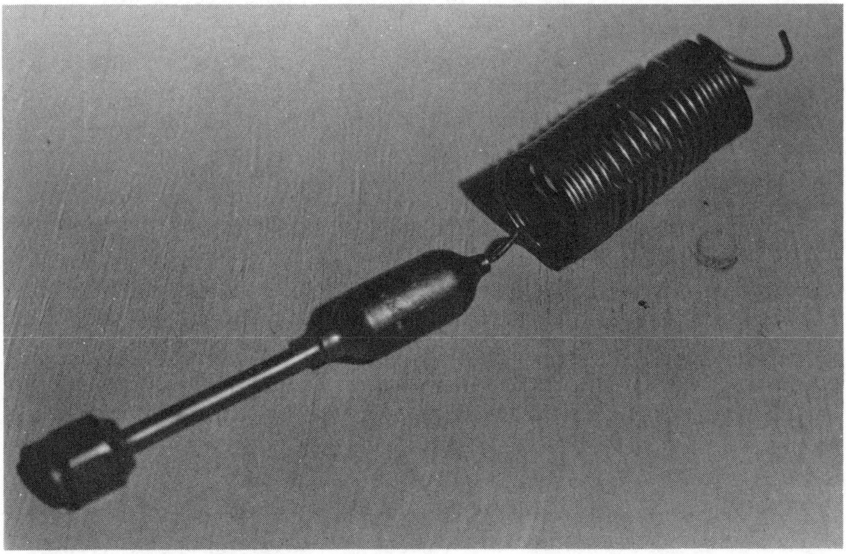

FIG. 4-1. The simplicity and low cost of the capillary tube make it ideal for many small refrigeration applications.

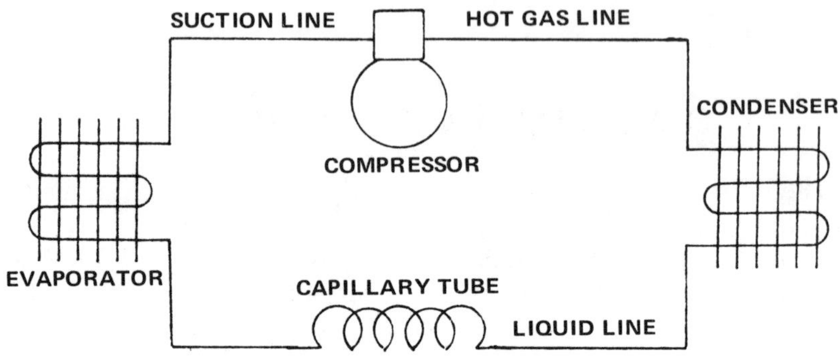

FIG. 4-2. Schematic of the cap tube system.

TXV function. A superheat sensing bulb generates a force, tending to open the valve. A spring and evaporator pressure tend to close the valve. When the compressor stops and evaporator pressure increases, what happens to the valve? It closes. This simple closing of the TXV valve, during the off cycle, is the main difference between the TXV and the cap tube.

Since the TXV is closed, the pressure in the evaporator and the pressure in the condenser are slow to equalize and we need a high starting torque compressor to start against the high pressure maintained in the condenser. A cap tube system, on the other hand, will equalize the two pressures rather quickly. Within a few minutes, the high and low side pressures are equal, hence the lower starting torque requirement.

This rapid equalization of the two pressures means that only thermostats can be used for defrost control. If a pressurestat were used, it would soon read the average of the evaporator and condenser pressures and cut in before the coil was completely defrosted.

Figure 4-2 shows a sketch of a capillary tube system. The cap tube itself is a very small diameter tube, the bore ranging from .026″ to about .055″ dia. The function of any refrigerant flow control is to provide a restriction against which the compressor can pump, thereby developing an elevated high pressure, and to meter the refrigerant into the evaporator so that a low pressure, hence low temperature, is achieved, utilizing the evaporator surface to a maximum.

The TXV worked admirably in fulfilling its intended purpose. As superheat in the evaporator coil varied, the valve responded, feeding in enough refrigerant to match the required cooling load. The cap tube has no moving parts. It is designed to work at maximum efficiency at only one combination of high side and low side pressures. Oddly enough, though handicapped in this way, the cap tube responds rather nicely over a surprisingly wide range of operating conditions. It works so well, in fact, that it is used universally in domestic applications and has made inroads in smaller commercial applications.

FUNCTIONING

Figure 4-3 shows a capillary tube system and has identified a point somewhere along the length of the cap tube as the *bubble point*. It is at this point that liquid refrigerant begins to boil off into a vapor. The state of the refrigerant upstream of this point is liquid and the state of the refrigerant downstream of this point is a mixture of liquid and vapor.

There is greater resistance to flow of refrigerant in that part of the tube that has a mixture of liquid and vapor. Therefore, the closer the bubble point is to the condenser side of the cap tube, the greater will be the resistance to flow. As a result, there will be less refrigerant flowing and system capacity will drop.

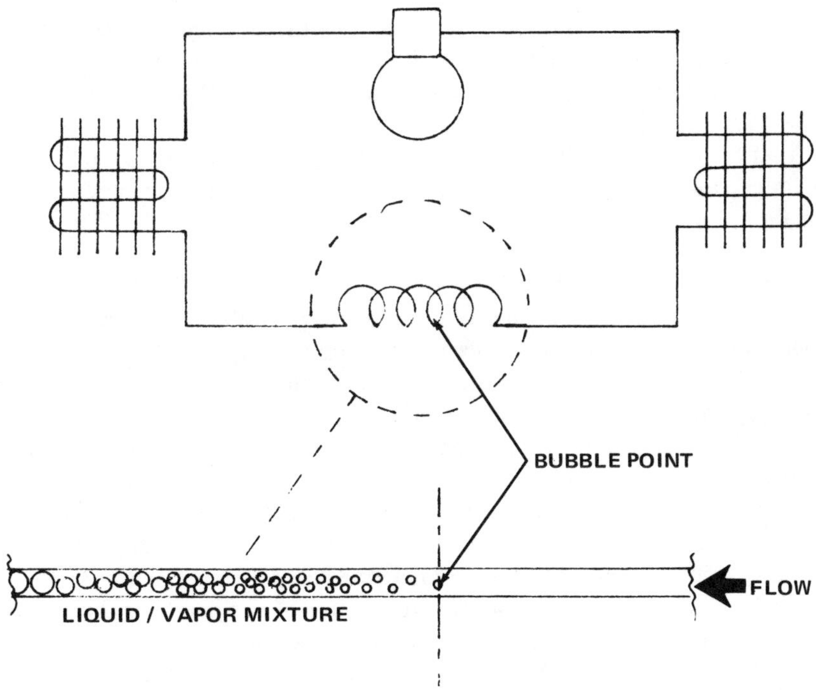

FIG. 4-3. Bubble point in the cap tube system.

Conversely, as the bubble point approaches the evaporator inlet, there will be less resistance to refrigerant flow, more refrigerant flowing and an increase in system capacity.

The location of the bubble point is not a random event, but directly related to the operating conditions of the system. If the refrigerator is subjected to a high load, that is, warm air flows over the evaporator as a result of increased product loading or more door openings, the low side pressure will increase because more refrigerant is boiling off in the evaporator. At higher pressures, the refrigerant is more dense and the compressor more efficient. As a result, more refrigerant is pumped into the condenser.

This increased refrigerant flow causes liquid to back up in the condenser, increasing high side pressure and subcooling of the liquid. The combination of higher pressure and greater subcooling causes the bubble point to move toward the evaporator. Since it is desirable to have refrigerant back up in the condenser during high load operation, cap tube systems should not have receivers.

Under low road operation, fewer pounds of refrigerant vapor are generated in the evaporator and lower compressor efficiency results. As a consequence, less liquid is condensed and there is less subcooling and lower pressure in the condenser. Under these conditions, the bubble point moves back toward the condenser, thereby reducing refrigerant flow and cooling capacity.

You can see that, while the capillary tube has no moving parts, system changes cause variations in refrigerant flow and a meaure of control is obtained.

OTHER CAP TUBE CONSIDERATIONS

To assure that all cap tube units of a given model will have identical operating characteristics, the capillary tubes must be identical in three ways: they must have the same bore, they must have the same length, and they must be installed in the same way.

The first and second requirements are easily duplicated, but the third requirement can become challenging. Some capillary tubes are coiled, some are installed straight, some are soldered to the suction line, and some are merely strapped to the suction line. On a production line, the manufacturer is able to produce identical cap tube installations. In the field, however, the serviceman must be aware of the need for duplicating the manufacturers' installation as closely as possible.

Soldering or strapping the cap tube to the suction line, Figure 4-4, increases subcooling. As we've seen, this will tend to keep the bubble point close to the evaporator, resulting in higher capacities. Replacing a

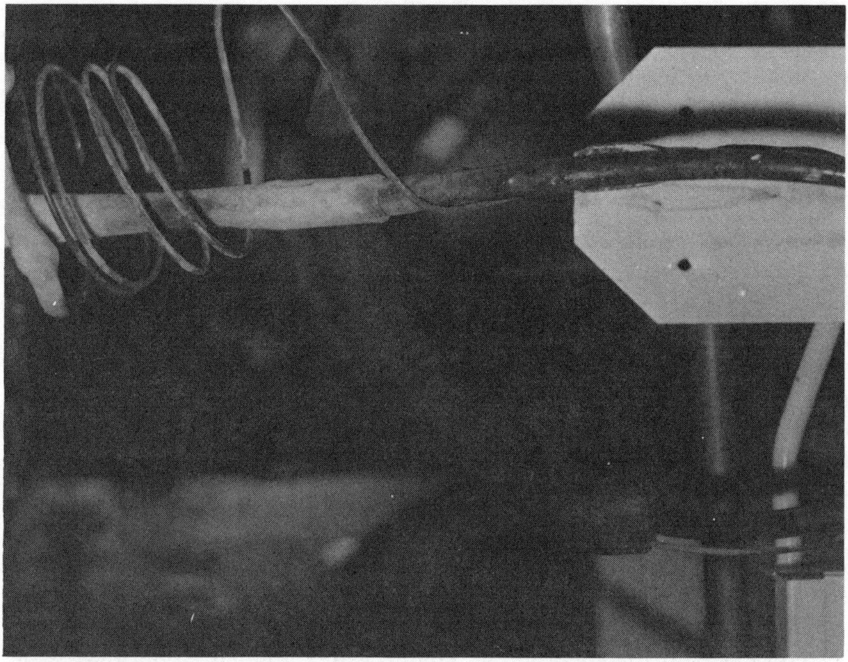

FIG. 4-4. This cap tube is wrapped around the suction line to increase subcooling. Note the loose coiling of the tube just before it enters the evaporator area.

soldered cap tube with an unsoldered one could have a definite effect on the system's performance.

Coiling the capillary tube produces a specific resistance to flow. Not coiling a replacement cap tube, or coiling it in a different manner, may produce a different flow resistance and alter system capacity.

Of greater importance, in field replacement, is the requirement that a clean cut be made when shortening a cap tube to the proper length. A burr at the end of a cap tube will act as a restrictor and severely inhibit proper performance.

The major problem with capillary tubes is a blockage. The blockage usually has one of two sources: a small particle of solid matter, such as a bit of dirt, a copper chip, or some such, may become lodged in the tube, partially or totally blocking it; or free water may find its way into the refrigerant stream and freeze in the cap tube.

When a blockage is encountered, first treat it as a freeze-up, much the same way a freeze-up in a TXV system was handled. If this doesn't work, assume a solid particle has permanently restricted the flow of refrigerant.

While there are various types of devices for hydraulically clearing a blockage, perhaps the best procedure is to remove and replace the entire cap tube. If a fine mesh strainer is in the circuit, replace it; if not, install one. This is not a drier, but merely a mechanical filter to prevent any other solid particles from getting into that cap tube.

TEMPERATURE CONTROL

Since the cap tube does not close during the off cycle, a pressure control, to maintain temperature, is not feasible. Rapid equalization of system pressures will result in continual short cycling of the compressor.

Thermostats are used for temperature control, operating as they would with TXV systems. The sensing element of the thermostat can be mounted on the coil surface where it senses surface temperature. As we've seen, the definite, reproducible relationship between the refrigerant, hence surface temperature, and the air temperature in the box, permit this control to shut down the compressor at any selected temperature.

Since defrosting of the coil is necessary, the cut-in temperature of this thermostat must be in excess of 32°F to assure that the ice has had sufficient time to melt.

The sensing bulb is usually placed on the suction line or imbedded in the coil fins near the suction line. Under a high load situation, the refrigerant vapor will be quite warm by the time it reaches the suction line. The sensing bulb, being warm, would keep the compressor running until the box became colder and the superheat decreased enough so that the sensing bulb would get cold and shut down the compressor.

If there was a shortage of refrigerant in the system, the superheat might never decrease enough to affect the sensing bulb. The compressor will run continuously, ice would build up to prodigious proportions and the box will get warm.

This is one reason why the refrigerant charge of a cap tube system is critical. Too much or too little refrigerant will cause erratic temperature control.

Another reason for the refrigerant charge being critical goes back to the explanation of bubble point. If excessive refrigerant is present in the condenser, higher pressures and subcooling will result in more refrigerant passing through the cap tube. If, as a result of adding too much refrigerant to an undercharged unit, high side pressures build up excessively, reduced capacity and compressor damage could occur.

REFRIGERANT CHARGE

While most larger commercial units are equipped with one or more service valves, it is not unusual to find that smaller, hermetic compressors

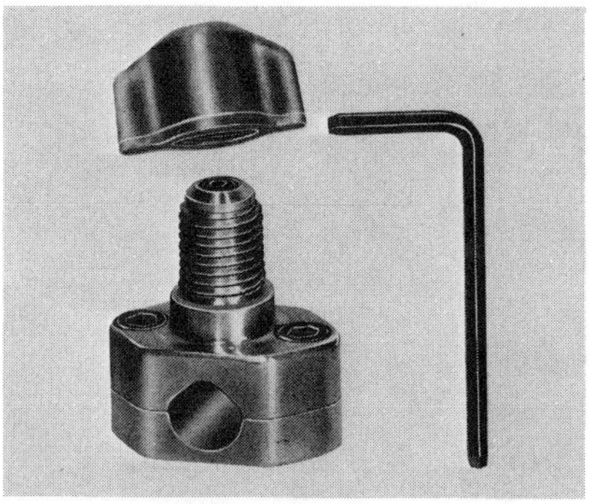

FIG. 4-5. Line tap valve provides quick access to the refrigerator system. *Courtesy Watso, Inc.*

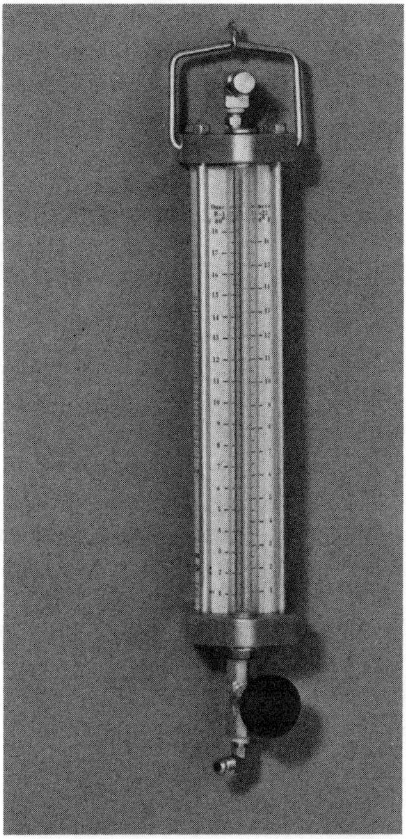

FIG. 4-6. A charging cylinder measures refrigerant in ounces. Note the scales for various refrigerants. *Courtesy Thermal Engineering Co., Div. Seakay Co., Inc.*

are not so equipped. The line tap valve, Figure 4-5, provides quick access to the system. It is designed to pierce the refrigerant lines, thus opening the system for pressure checks and for recharging. On older systems, where the service valves are corroded, the line tap valve may be used to regain access to the system.

It is rather difficult to add just enough refrigerant to an undercharged cap tube system. Since cap tube systems have no receivers and are generally found on smaller sized units, it is usually no hardship to release the remaining refrigerant, evacuate the system with a vacuum pump, and recharge with the amount of refrigerant specified on the unit nameplate. If you use a charging cylinder, Figure 4-6, you can measure the refrigerant in ounces.

If the amount of refrigerant in the system is unknown, the *frost line technique* is useful for getting a fairly good approximation of the correct charge. Refrigerant is added until frost is observed on the suction line back to the compressor. Refrigerant is then released gradually from the system and the *frost line* on the suction line is observed. When the frost line has backed off, to about 2 to 3 feet from the compressor, an adequate charge is present.

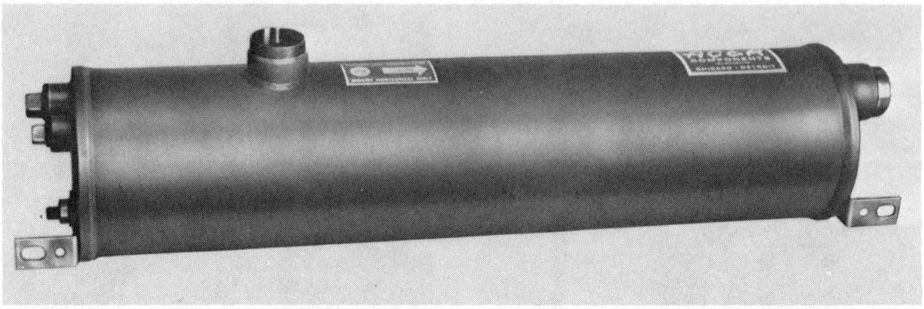

FIG. 4-7. Suction accumulator. *Courtesy A C & R Components, Inc.*

The determination of an adequate charge should only be made when the box is close to its normal operating temperature. If the box is too warm, a frost line 2 to 3 feet from the compressor could move up to the compressor as the box temperature continues to decrease. Liquid refrigerant might then get into the compressor with dire consequences. This charging procedure, if properly done, is rather time consuming and evacuation and charging by weight is recommended.

While the cap tube does have a certain ability to meter refrigerant flow, there will be times, particularly under low load conditions, when too much liquid refrigerant enters the evaporator. The suction accumulator, Figure 4-7, is a device installed in the suction line to provide a place for this excess liquid to accumulate. In time, the liquid in the accumulator will boil off into a vapor and the vapor will be drawn back to the compressor. While some elaborate accumulators are available, the usual accumulator is no more than a bulb-shaped expansion of a portion of the suction line, similar in appearance to a strainer.

We're going to take a closer look at hermetic compressors, particularly their wiring diagrams, starting components, and other electrical characteristics. Before we do that however, we'll take a closer look at the vapor compression cycle. A thorough understanding of this cycle may make many of the things previously stated fall into place a bit better.

5

The Mollier Diagram

The Mollier diagram, or pressure-enthalpy diagram, as it is sometimes called, presents the thermodynamic properties of refrigerants in diagrammatic form. Figure 5-1 is the Mollier diagram for R-12. It is a horrendous looking affair and, unless you have some experience working with graphs of this sort, the appearance of all those lines will boggle the mind. The Mollier diagram is a basic tool in refrigeration system analysis and, as such, understanding it provides some good insights into what is happening in a system and the reason for it.

The most prominent feature of the diagram is the dome shown in Figure 5-2. The dome actually is made up of two curved lines, the *saturated liquid* line and the *saturated vapor* line. These lines meet at the *critical point*. This point is defined by the *critical temperature*, a tem-

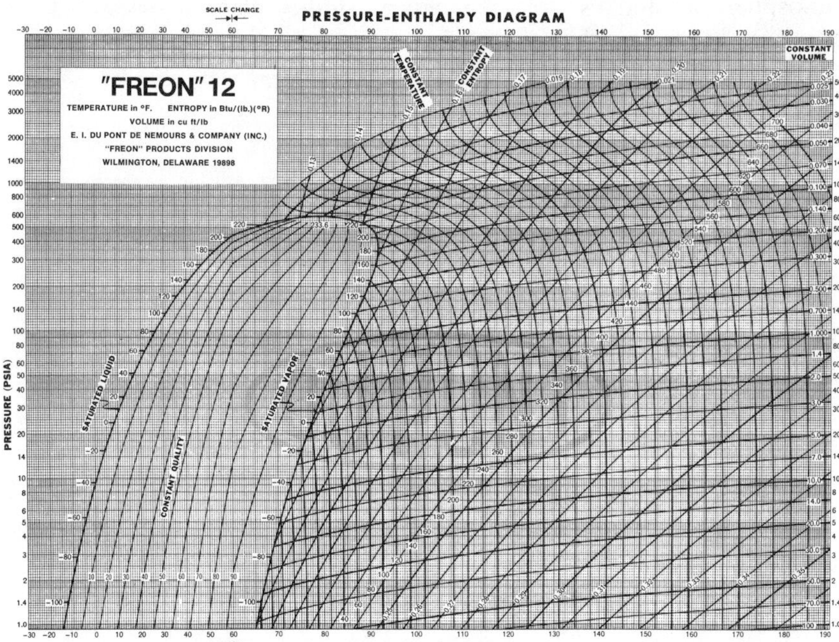

FIG. 5-1. Mollier diagram or pressure-enthalpy diagram for Refrigerant-12. *Courtesy E. I. duPont de Nemours & Company.*

perature above which the refrigerant will never condense, no matter how high the pressure, and the *critical pressure,* which is the pressure corresponding to the critical temperature on the saturation curve.

The state of the refrigerant is defined by the limits of the dome. To the right of the right-hand line, or saturated vapor line, the refrigerant is superheated. That is, the refrigerant is totally vapor, with no liquid present. To the left of the left-hand line, or saturated liquid line, the refrigerant is subcooled. That is, the refrigerant is totally liquid, with no vapor present. Inside the dome the refrigerant is in a saturated condition and is a mixture of vapor and liquid. In describing the vapor compression cycle, you'll remember that superheat, subcooling and vapor/liquid mixtures are involved. So far there is nothing new.

Figure 5-3 shows horizontal lines superimposed on the dome. The pressure along any one horizontal line is constant and can be determined by following that line over to the left-hand side of the diagram and reading the pressure. Note the pressure is psia, pounds per square inch absolute. The pressure we read on our pressure gages is psig, or pounds per square inch gage. To change from psig to psia, we merely add 15 to the psig reading, or 14.7 to be exact. This represents the pressure of the atmosphere around us. Just remember, Mollier diagram pressures are expressed in psia which is about 15 pounds greater than gage pressure.

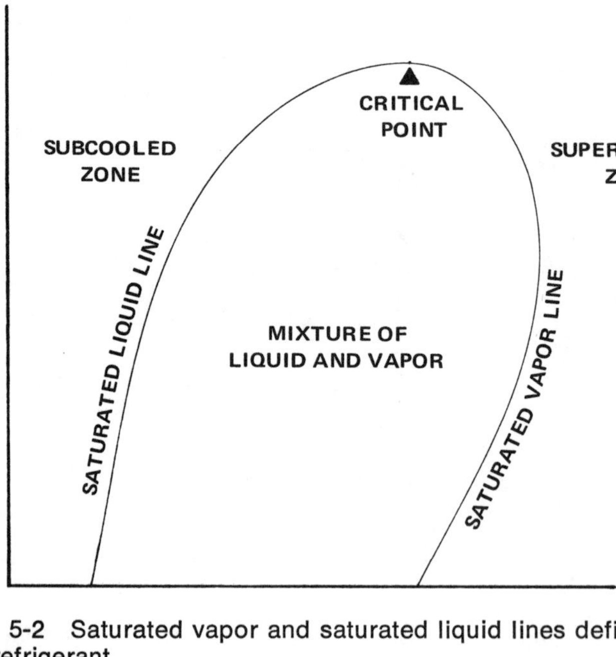

FIG. 5-2 Saturated vapor and saturated liquid lines define the state of the refrigerant.

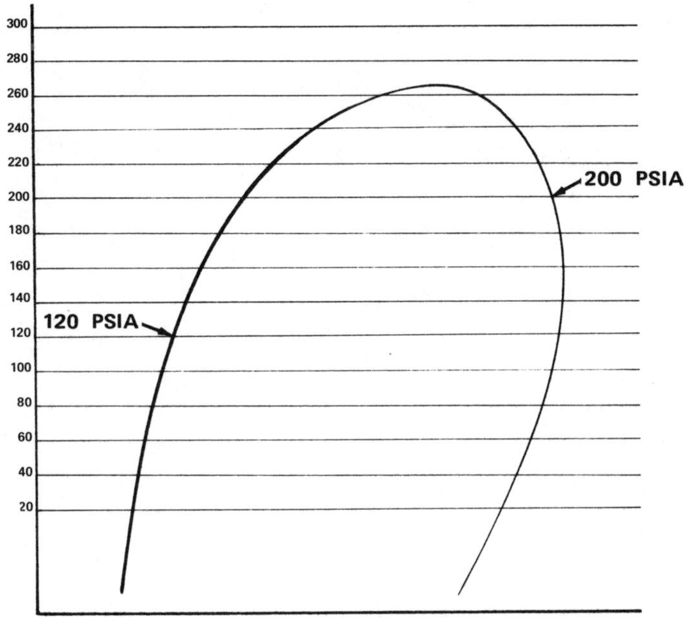

FIG. 5-3. Constant pressure lines.

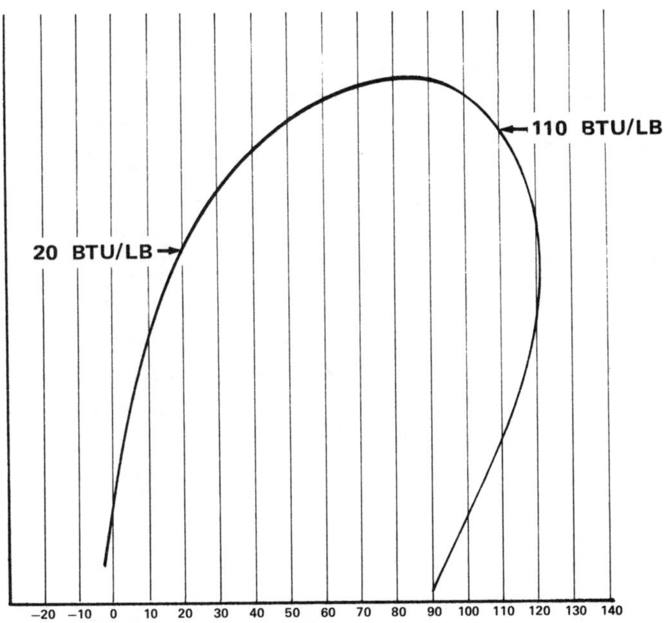

FIG. 5-4. Lines of constant enthalpy.

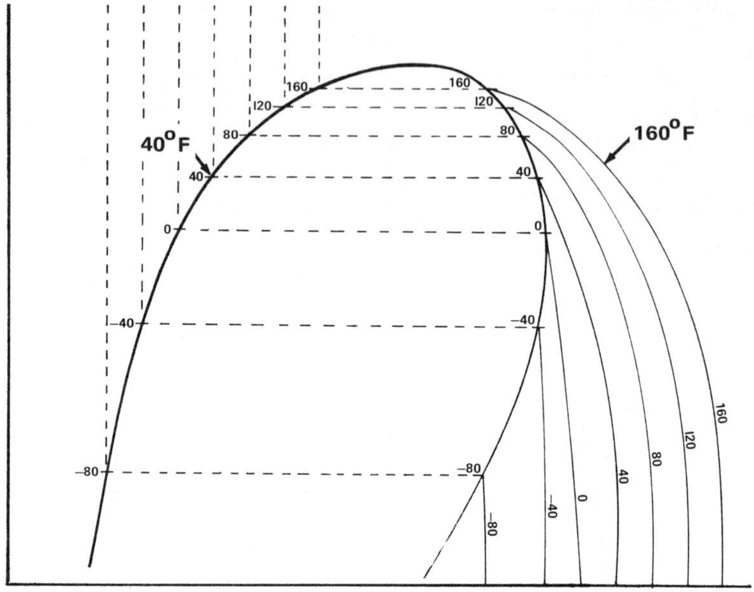

FIG. 5-5. Constant temperature lines.

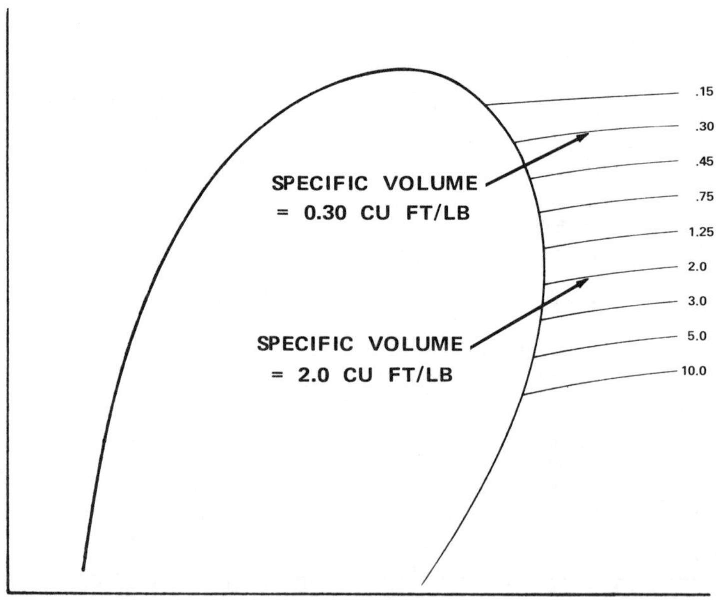

FIG. 5-6. Lines of constant specific volume.

In Figure 5-4 we've removed the pressure lines, to keep it simple, and superimposed lines of constant enthalpy. Enthalpy is a fancy word for heat content and is expressed in Btu/lb (British thermal units per pound of refrigerant). The heat content of the refrigerant at any point on the diagram can be found by following a vertical line downward to the bottom of the graph and reading enthalpy (heat content) in Btu/lb.

Figure 5-5 shows lines of constant temperature. Note that they differ from the other two lines we've looked at so far. The lines are vertical, in the subcooled zone, then become horizontal in the vapor/liquid zone, and curve away downward in the superheat zone. The dotted lines of Figure 5-5 are usually not shown on the Mollier diagram. Instead, they are represented by horizontal dashes at the liquid and vapor saturation lines, Figure 5-1.

The lines drawn in a flat arc from the saturated vapor line toward the right in Figure 5-6 are lines of constant specific volume. At a point along one of these lines you will find the number of cubic feet that one pound of refrigerant will occupy.

Lines of constant entropy are shown in Figure 5-7. Entropy is defined as the ratio of heat content in a pound of refrigerant vapor to its absolute temperature. The definition itself is not vital to service technicians, but

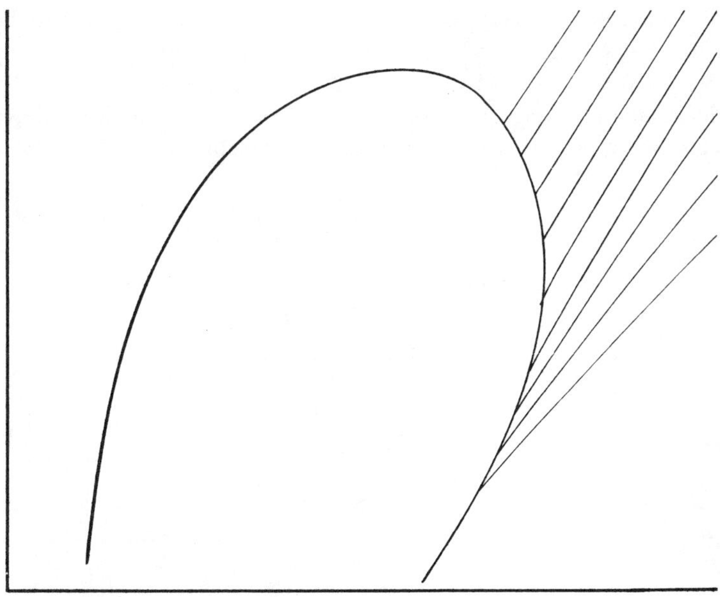

FIG. 5-7. Lines of constant entropy.

its significance is quite important. The compression of the refrigerant is assumed to be an isentropic (constant entropy) process. You will see the significance of this shortly.

There you have the five bits of thermodynamic information that are graphically presented by the Mollier diagram. Now let's put it all together and relate this information to what's happening in the mechanical vapor cycle.

READING THE MOLLIER DIAGRAM

A typical reach-in refrigerator capillary tube system is shown in Figure 5-8. The Mollier diagram for R-12, showing the properties of the refrigerant as it flows through the system, is shown in Figure 5-9.

Let's start at point W in the cycle. A mixture of vapor/liquid is absorbing heat in the evaporator. The enthalpy (heat content) at W is 28.0 Btu/lb. As the refrigerant flows through the evaporator, more and more liquid changes to vapor until at point X all the liquid has changed to vapor and the heat content of the refrigerant is 79.0 Btu/lb. The vapor continues to flow through the evaporator, picking up a bit more heat, until it reaches point Y at 30°. The heat content is now 81 Btu/lb. The heat that the refrigerant picked up in the evaporator is found by subtracting the heat content of the refrigerant when it entered the evaporator at W from the heat content upon leaving the evaporator ($81 - 28 = 53$ Btu/lb). Therefore, every pound of refrigerant passing through this evaporator has picked up 53 Btu's of heat from the refrigerator compartment. This value is called the *net refrigerating effect.*

Notice that the refrigerant, as it flows through the suction line, picks up a bit more superheat, its temperature increasing to 33°F at the com-

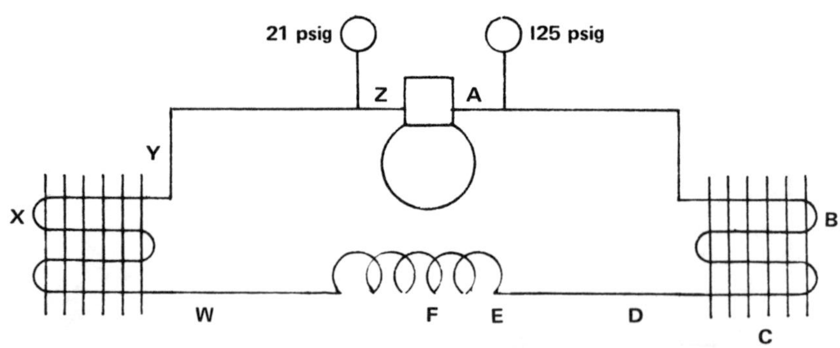

FIG. 5-8. Capillary tube system.

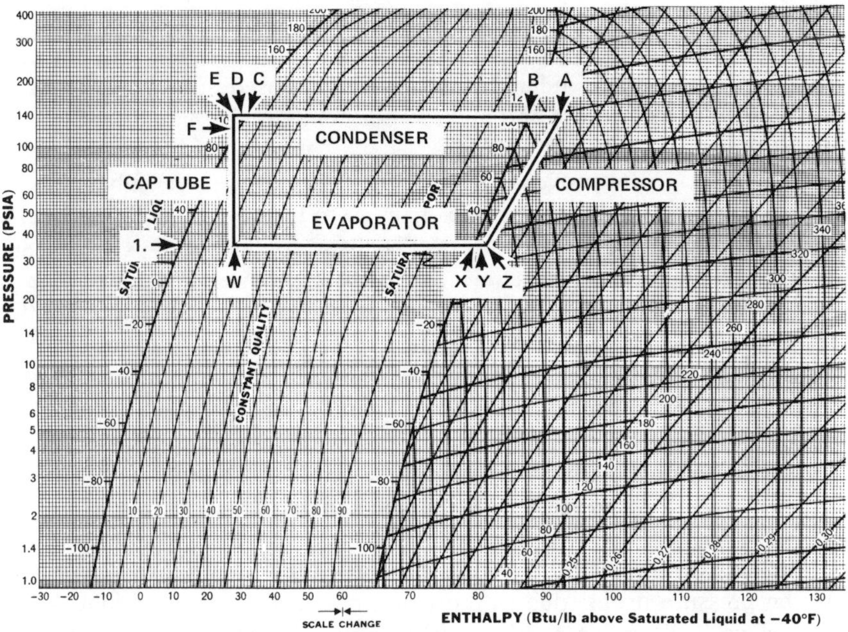

FIG. 5-9. Mollier diagram of R-12 as it passes through a capillary tube system.

pressor. This extra heat was not absorbed in the evaporator and therefore performs no useful cooling and does not increase the *net refrigerating effect.*

Notice that the pressure from W to Z is constant, that the temperature from W to X is constant, but then increases from X to Z, indicating that superheating is taking place.

The refrigerant enters the compressor at Z and leaves the compressor at A. The line from Z to A is a compression line and you'll recall that compression is a constant entropy or isentropic process. Therefore, the compression line is drawn parallel to a line of constant entropy from point Z to point A which, being 140 psia, is the high side pressure of the system. Remember, Mollier diagram pressures are in pounds per square inch *absolute.*

The heat content at point Z was 81.5 Btu/lb and is seen to be 91 Btu/lb at point A. Therefore, each pound of refrigerant absorbed 9.5 Btu during the compression process. This is called the *heat of compression* and can be interpreted as representing the amount of work the compressor must do to compress the refrigerant since heat is energy and work is energy.

The refrigerant now moves from the compressor into the condenser. Note that the compressor discharge temperature is 128°F. In moving from A to B in the condenser, the refrigerant is desuperheated. As it moves from B to C, more and more vapor condenses to liquid until, at C, all the vapor has changed to liquid.

As the refrigerant moves from C to D in the condenser, subcooling is taking place and finally, as it moves from D to E through the liquid line, a bit more subcooling has occurred. The heat content at E is now 28.0 Btu/lb. The amount of heat rejected in the condenser and liquid line is the heat content at A less the heat content at E or 91 − 28 = 63 Btu/lb. The heat rejected is greater than the heat absorbed in the evaporator. This fact explains why condensers are usually larger than evaporators and why they usually have more air passing over them.

Heat Absorbed in the Evaporator + Heat of Compression = Heat Rejected in the Condenser.

The refrigerant then enters the capillary tube as a subcooled liquid. As the pressure drops, due to the resistance to flow, the subcooling decreases until, at point F, the *bubble point,* liquid starts flashing into vapor. As the mixture of vapor and liquid passes through the cap tube, the pressure continues to drop until the refrigerant enters the evaporator, at point W, at a pressure of 36 psia. The temperature dropped from 92°F to 20°F, yet the enthalpy or heat content remained unchanged.

If we could introduce the liquid refrigerant into the evaporator as 100% liquid, that is, if we could totally eliminate flash gas, we would be at point 1. The heat content of the refrigerant would be 13 Btu/lb. The amount of heat that this liquid could absorb, if it changed completely to vapor, is represented by a line between point 1 and point X. This is 71 − 13 or 66 Btu/lb. The *actual* heat absorbed in the cycle is represented by the line from W to X. This is 79 − 28 or 51 Btu/lb. Therefore 66 − 51 or 15 Btu/lb is lost in flash gas. This can be calculated as a percentage:

$$\frac{15}{66} \times 100 = 23\%$$

This percentage can also be obtained by observing the position of point W with reference to the constant quality lines shown under the dome. These lines give the percentage of vapor in the vapor/liquid mixture of refrigerant. Notice that, as we follow line W - X toward the right, the amount of vapor in the mixture increases toward 100%.

The cycle is now complete. Go through this a few more times, referring to both the chart and the diagram, to get a feeling for what is happening.

SYSTEM ANALYSIS WITH THE MOLLIER DIAGRAM

Comprehension of the Mollier diagram can help answer many questions about operation of the system under other than design conditions. Here are some examples:

WHAT IS THE EFFECT OF HIGH DISCHARGE PRESSURES?

Figure 5-10 shows both the diagram we've been looking at and, superimposed on it, a diagram representing a system with elevated discharge pressure. The original high side pressure of 140 psia has increased to 215 psia or about 200 psig. We can see that the quality of the refrigerant entering the evaporator has increased from 23% to about 38% vapor. This means that flash gas has increased by this amount and useful refrigeration has decreased. The heat of compression has increased from 9.5 Btu/lb to 96 − 81.5 or 15.5 Btu/lb. This represents a significant increase in the power required to run the compressor. The greater the

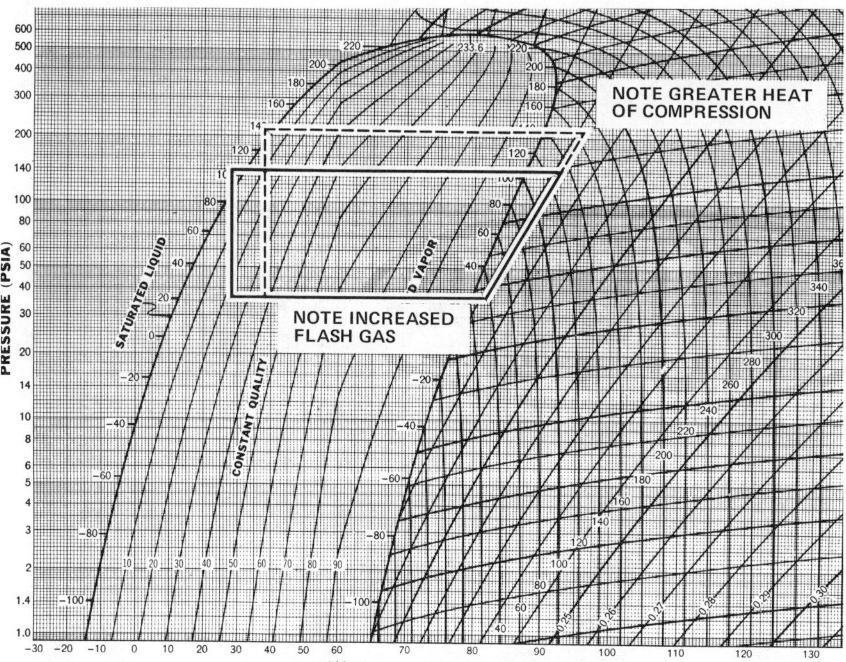

FIG. 5-10. Mollier diagram showing the effect of high discharge pressure on a cap tube system.

power requirement, the greater the current draw of the motor and the higher the motor winding temperature. The net result is shorter service life, with nuisance trip outs on overload. A third result is that the discharge temperature has increased from 128° to about 168°F. The lubricating effect of the oil is reduced at this higher temperature, hence more wear and tear on the motor. If the temperature gets high enough, a total breakdown of the oil structure can occur.

WHAT IS THE EFFECT OF LOW REFRIGERANT CHARGE?

With insufficient refrigerant, Figure 5-11, subcooling will decrease, resulting in more flash gas and lower refrigerating effect, therefore lower capacity. With less liquid in the evaporator, the refrigerant will boil off sooner, resulting in higher superheat at the compressor. Higher superheat reduces motor cooling by the refrigerant, leading to higher operating temperatures and early compressor motor failure. In addition, hotter vapor entering the compressor results in hotter vapor leaving the compressor, again heating the oil excessively and cutting down on lubrication capability.

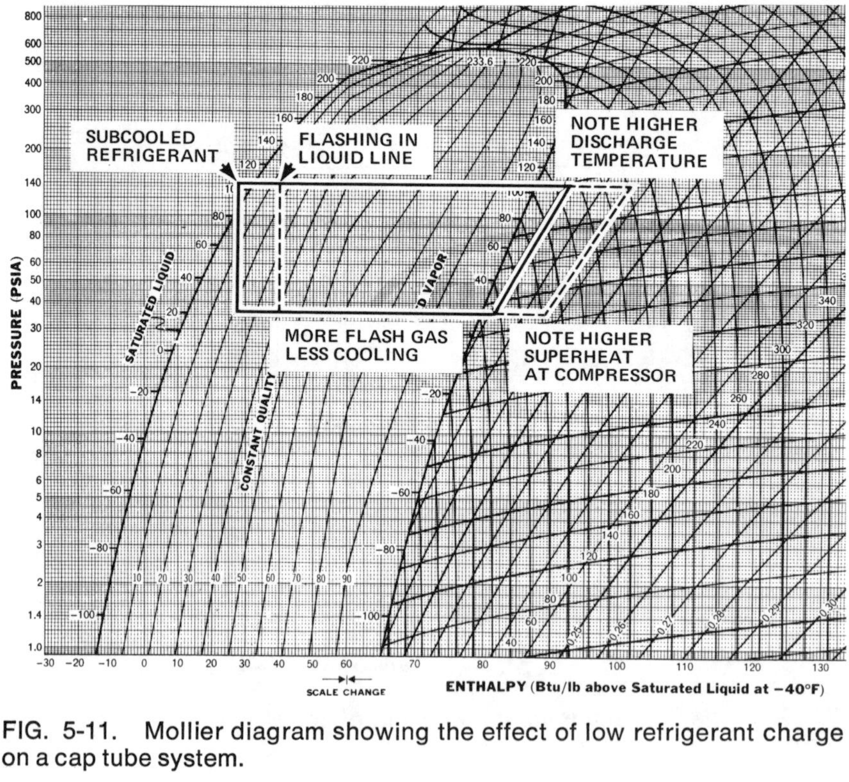

FIG. 5-11. Mollier diagram showing the effect of low refrigerant charge on a cap tube system.

WHAT IS THE EFFECT OF EXCESS REFRIGERANT IN THE SYSTEM?

Excess refrigerant in the system will cause liquid to back up in the condenser, resulting in greater subcooling which, at first glance, seems beneficial. Unfortunately, as the liquid builds up in the condenser, it causes higher discharge pressures. The effect of higher discharge pressure outweighs any benefit derived from more subcooling. The higher pressure forces more refrigerant through the capillary tube. This excess refrigerant may not all boil off in the evaporator, causing liquid to flood back to the compressor which will lead to dilution of the oil, hence reduced lubrication and possibly compressor failure, or the liquid refrigerant may enter the compressor cylinders and cause a mechanical failure.

WHAT IS THE EFFECT OF LOW DISCHARGE PRESSURE?

As discharge pressure drops, net refrigerating effect increases, therefore higher capacities are realized, Figure 5-12. The heat of compression will decrease, therefore operating current is less and efficiency is higher. Generally, lower head pressure results in more favorable operating characteristics. If the pressure gets too low, however, the force pushing refrigerant through the capillary tube decreases to the point where not enough refrigerant is being pumped. As an example, the 140 psia high

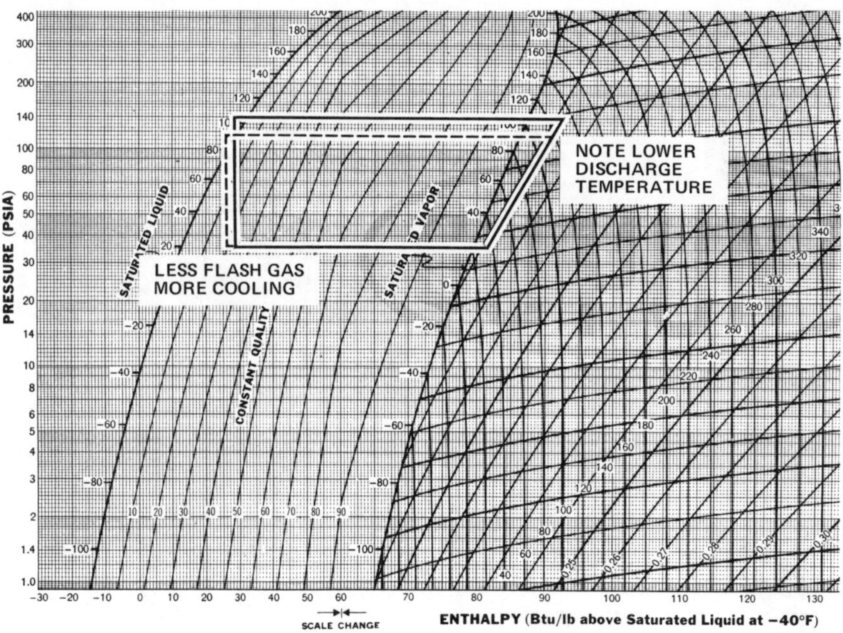

FIG. 5-12. Mollier diagram showing the effect of lower discharge pressure.

side and 36 psia low side have a pressure differential of 104 psi tending to push refrigerant through the cap tube. If the high side pressure dropped to 100 psia, the differential would only be 64 psi and refrigerant flow would decrease.

WHAT IS THE EFFECT OF LOW SUCTION PRESSURE?

Generally, the suction presure of the system is directly related to two things. First, the load on the evaporator. As the cooling load decreases, less vapor is generated in the evaporator and the pressure drops. Second, discharge pressure. As compressor efficiency increases, due to increased pumping capacity which accompanies decreasing head pressure, pressure will fall again.

The Mollier diagram shows that, at lower pressure, flash gas increases and net refrigerating effect decreases, Figure 5-13. Heat of compression, hence work, increases and efficiency drops off. The specific volume of the refrigerant, ft³/lb, also increases so that, even though the compressor is pumping the same volume of refrigerant, the weight of the refrigerant is less. The fewer pounds of refrigerant circulating, the lower the capacity.

Low pressures can be avoided by preventing high side pressures from getting too low. This is generally not a problem with reach-in refrigera-

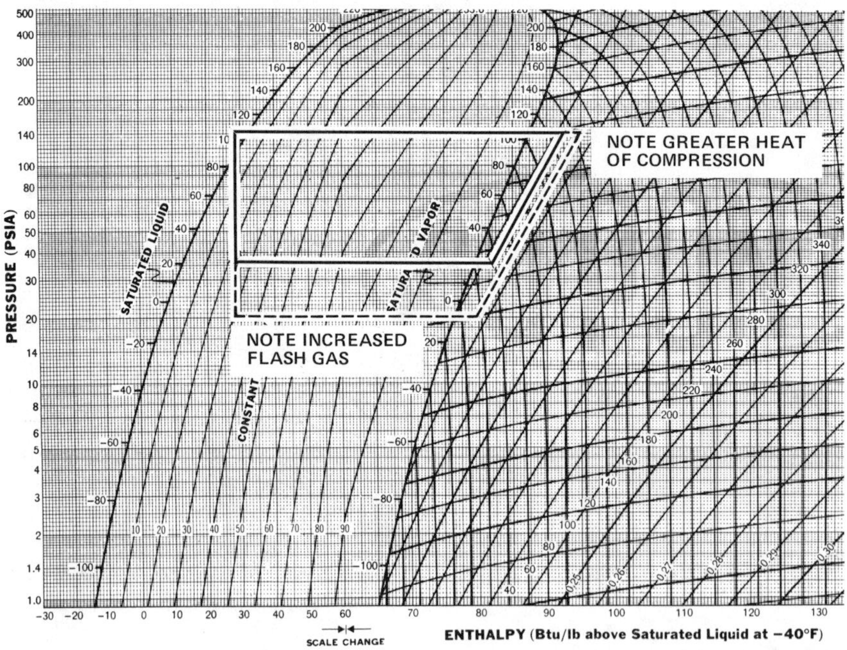

FIG. 5-13. Mollier diagram showing the effect of lower suction pressure.

tors although, if they are in low ambient locations, particularly in winter, they might not function as well as they should. This becomes more of a problem with remote systems, such as walk-in refrigerators. We'll see more of that later.

Low pressures can also be avoided by keeping the evaporator coil clean to insure maximum heat transfer and by making sure the evaporator fans are well lubricated and in good operating condition.

WHAT IS THE EFFECT OF HIGH SUCTION PRESSURE?

Actually high suction pressure is rather beneficial. Heat of compression is low, flash gas is low, efficiency is high and, since specific volume is low, more pounds of refrigerant are pumped, Figure 5-14. We are not only getting more cooling per pound of refrigerant, we are also getting more pounds circulating. A properly operating unit will have the highest suction pressure possible while still having a cold enough coil to do the cooling job.

On starting a refrigerator, the temperatures in the box will be quite high, generally around 70 to 80°F. This warm air puts a heavy load on the system, and high suction presures are quite common. As the box

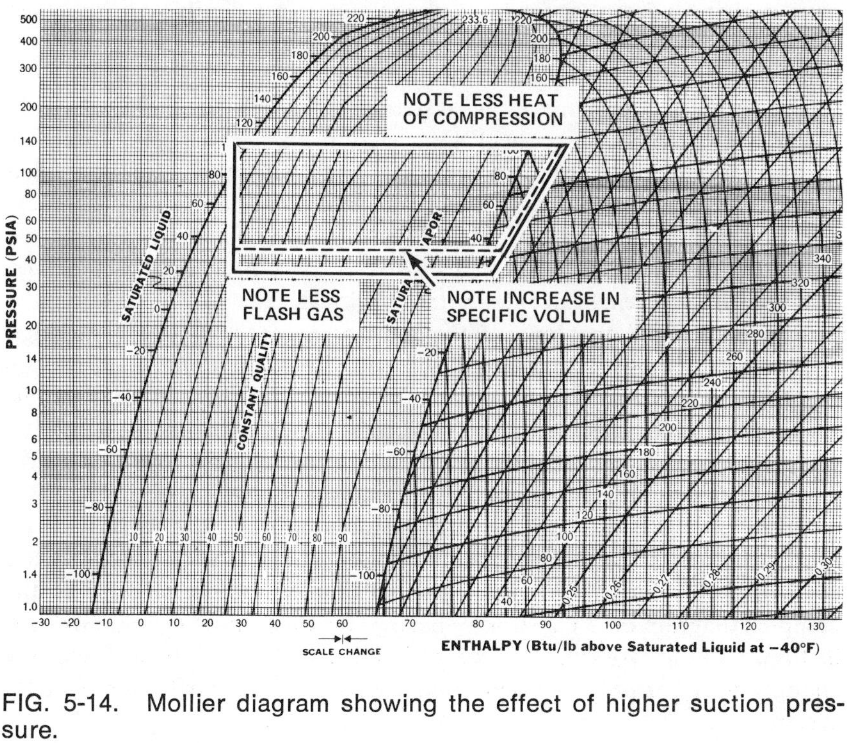

FIG. 5-14. Mollier diagram showing the effect of higher suction pressure.

temperature drops toward its normal operating level, the suction pressure and running current will gradually decrease to a normal operating condition.

It's interesting that many complaints of malfunctions are prefaced by the statement that the box had been off for some time, possibly because it was in storage, or was defrosted for a few days. Then, shortly after start-up, it failed. This really makes sense in light of what the Mollier diagram reveals. On start-up, with high box temperatures, the unit works very hard to bring the temperature down to normal. It is during this pull-down time that any structural or electrical weaknesses in the compressor may be aggravated and result in a major failure.

THE MOLLIER DIAGRAM AND TXV SYSTEMS

The capillary tube system is quite straightforward when laid out on the Mollier diagram. The TXV system is similar, except for one difference: TXV systems invariably have a receiver. Any subcooling developed in the condenser is lost as the refrigerant enters the receiver. The liquid leaving the receiver will be subcooled somewhat as it flows through the liquid line. The net effect is that the refrigerant entering the TXV will have perhaps a few degrees less subcooling than a similar cap tube system. Other than that, the analysis of the diagram is the same with both systems.

The result of overcharging the system is essentially the same with TXV or cap tube systems. Since the TXV system has a receiver, high discharge pressures, due to overcharging, will occur only after the receiver is filled. In a cap tube system, several ounces in excess of the design charge can result in erratic performance. In a TXV system, similar problems will occur only if several pounds of surplus refrigerant are added. The charge of a cap tube system is decidedly more critical.

You may recall that one of the reasons a cap tube can respond to changing conditions is that liquid will back up in the condenser, causing higher pressures and forcing more refrigerant into the evaporator. This very action precludes the use of a receiver in a cap tube system. If a receiver was present, the refrigerant would never back up in the condenser, but would accumulate in the receiver.

Fortunately for us in the service business, most of the work with the Mollier diagram has already been done. Tables, cards and shortcut *rules-of-thumb* have been developed so that we need not refer to this complex diagram on a day-to-day basis. This diagram is useful, however, in that, without working out complicated calculations, we can see how changes fom normal or design conditions affect the performance of the system.

6
Electrical Systems

The reach-in refrigerator is probably the simplest of all commercial refrigeration systems, both as a refrigeration system and electrically. Figure 6-1 shows an electrical schematic of a typical reach-in refrigerator. As soon as it is plugged in, the evaporator fan begins to run. It will run constantly from that moment. A light bulb operates in response to a door switch. The pressure control, which is in the system to control box tem-

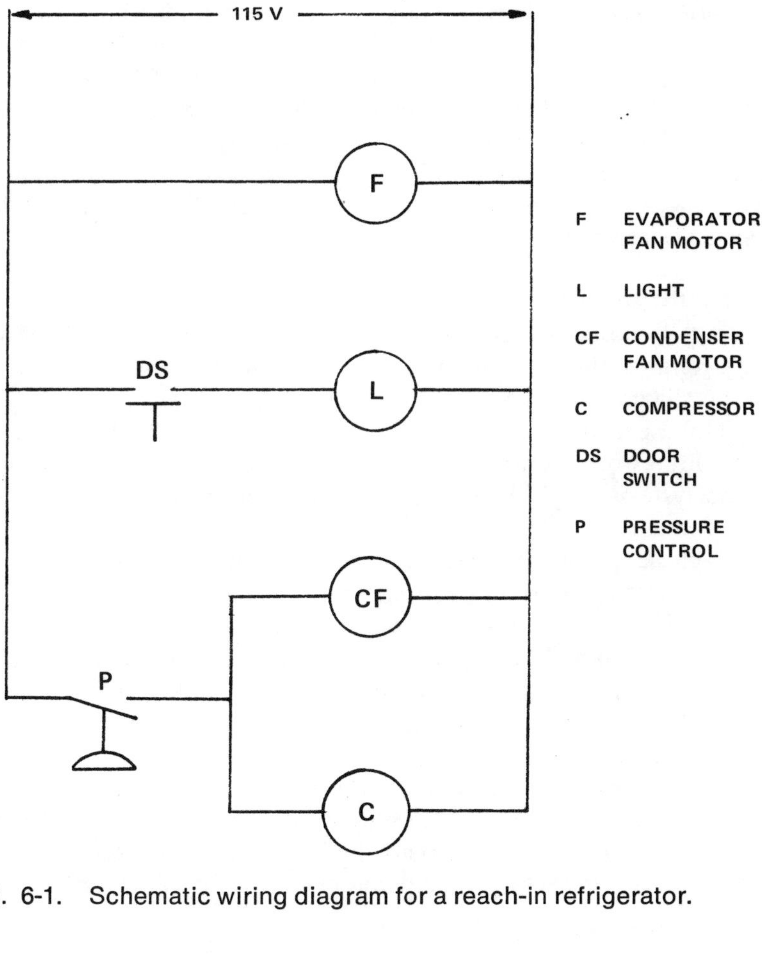

FIG. 6-1. Schematic wiring diagram for a reach-in refrigerator.

perature, controls the operation of both the condenser fan and compressor. When the control setting is reached, both the fan and compressor are de-energized.

This is a simple diagram. They get more complex from this point, so it would be a good idea to review wiring diagrams in general in order to form a base for the more complicated ones to come.

A wiring diagram is a representation of one or more electrical circuits. The purpose of the diagram is to enable a service technician to see how the various electrical components are related without having to pull apart the actual equipment and trace each wire. Two types of diagrams are generally used, pictorial and schematic. The pictorial diagram shows the location of the various components, identifies them, and shows, in fairly good detail, how they are wired together. The schematic shows how they function electrically and is usually more helpful in troubleshooting than the pictorial, although having both available is paradise.

Many of the smaller commercial units use 115 volt, single-phase power. Large reach-in units and many walk-in boxes use 230 volt, single-phase power. There is not too much difference in the diagrams or in the troubleshooting techniques used for either power supply. The 208/220 volt, three-phase equipment used with larger walk-in systems is somewhat different and we will look at that in a later chapter. For now we will deal with single-phase systems.

COMPRESSOR WIRING

Figure 6-1 shows only two wires going to the compressor. Many unit wiring diagrams are simplified in this way, with the compressor having a separate diagram to show how it is wired.

Single-phase compressors can be classified as low or high starting torque compressors. Low starting torque compressors are used on capillary tube systems only. High starting torque compressors are used with TXV systems.

Single-phase motors, by their very nature, require an added electrical winding for starting purposes. Three-phase motors do not have this requirement. This auxiliary winding is called a start winding. Once the motor has started and has attained near normal speed, there is no further need for the start winding. The winding that keeps the motor running, once it has come up to operating speed, is called the *run* winding or *main* winding.

The compressor wiring diagram will show the start winding, the run winding, a device used to take the start winding out of the circuit when it is no longer needed, called a relay, and a safety device used to sense any over-current or excessive temperatures caused by an overload.

LOW STARTING TORQUE COMPRESSORS

A low starting torque compressor is shown in Figure 6-2. This type of compressor is generally found on smaller reach-in boxes and other small commercial systems such as water coolers, sandwich tables, ice makers, etc.

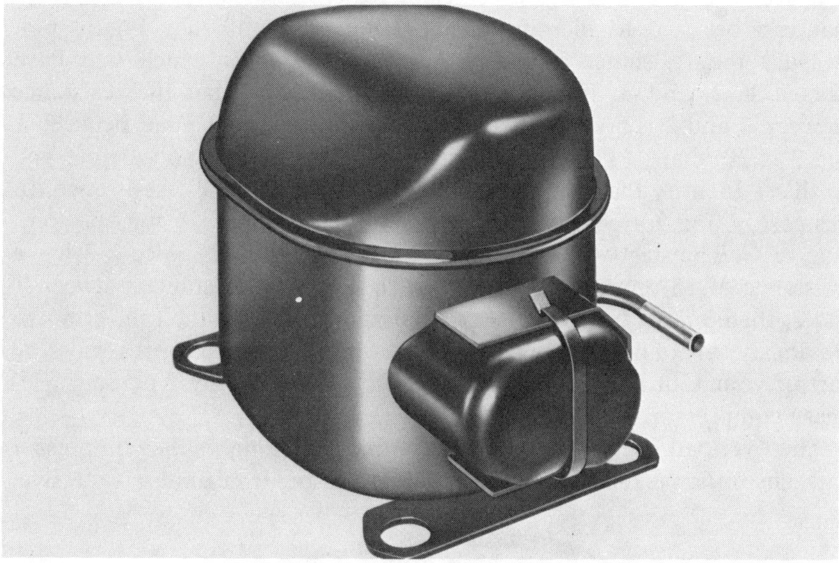

FIG. 6-2. Low starting torque compressor used with cap tube systems. The compressor terminals and starting relay are under the cover at the side of the compressor. *Courtesy Tecumseh Products Co.*

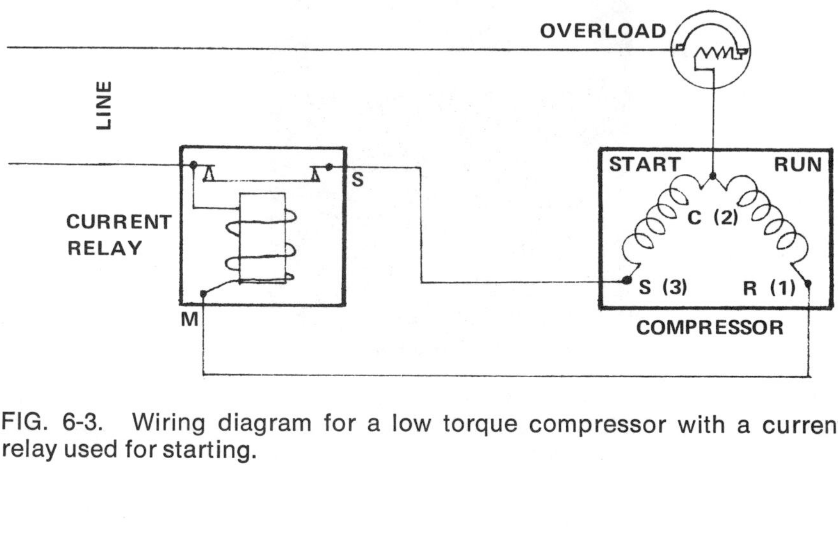

FIG. 6-3. Wiring diagram for a low torque compressor with a current relay used for starting.

The wiring schematic for low torque compressors is shown in Figure 6-3. The run winding of the compressor is between C-R and the start winding is between C-S. Terminal C will then be the common terminal, R the run winding terminal, and S the start winding terminal. This coding is practically universal.

If the terminal markings are illegible, there is an easy electrical test that can be used to identify them. Using an ohmmeter, Figure 6-4, measure the resistance between the three terminals which you have labeled 1, 2, and 3. For example: you might find that the resistance between 1 and 2 is 5 ohms, between 2 and 3 is 15 ohms, and between 1 and 3 is 20 ohms. (To be safe, disconnect the power supply before you do this.) Identify the terminals that registered the highest resistance. In this case, 1 and 3 are R and S. The remaining terminal, 2, must be common or C. The start winding has the greatest resistance. Since 2-3 has a resistance of 15 ohms, compared with the 5 ohms resistance between 1 and 2, then 3 must be the start terminal and 1 must be the run terminal. Obviously, an ohmmeter capable of reading small resistances must be part of your tool kit. This procedure is identical for any type of single-phase compressor.

The overload is connected to the common winding of the compressor and senses the current flowing to the compressor. If current is excessive,

FIG. 6-4. Inexpensive volt-ohmmeter. The resistance of most small compressors will be less than 20 ohms, which this meter will measure quite accurately. *Courtesy Triplett Corp.*

it will open the circuit and the unit will shut off. Since the overload is mounted on the body of the compressor, it also senses the temperature of the compressor housing. If the housing becomes too hot, possibly due to poor suction gas cooling, it will also open the circuit.

The starting relay may be one of two types, a current relay or a voltage relay. In this case, we have a current relay.

The current relay has three terminal markings: L for line, S for start and M for main or run. This also is a fairly common way of identifying these terminals, regardless of manufacturer. The relay has a coil and a switch built into a common housing. In the absence of any note to the contrary, you can identify a current relay by the terminal markings and the fact that the coil is visible, Figure 6-5.

The schematic, Figure 6-3, depicts the relay in the start position. At the instant of start-up, current flows through the relay coil and the run winding. The magnitude of this current is 4 or 5 times the normal running current of the compressor. This high current creates a magnetic force within the relay, pulling the switch contacts closed, allowing current to now flow to the start winding. The compressor motor starts turning and comes up to speed. As it approaches approximately 90% of full speed, the current drops toward normal running current. This lower current is

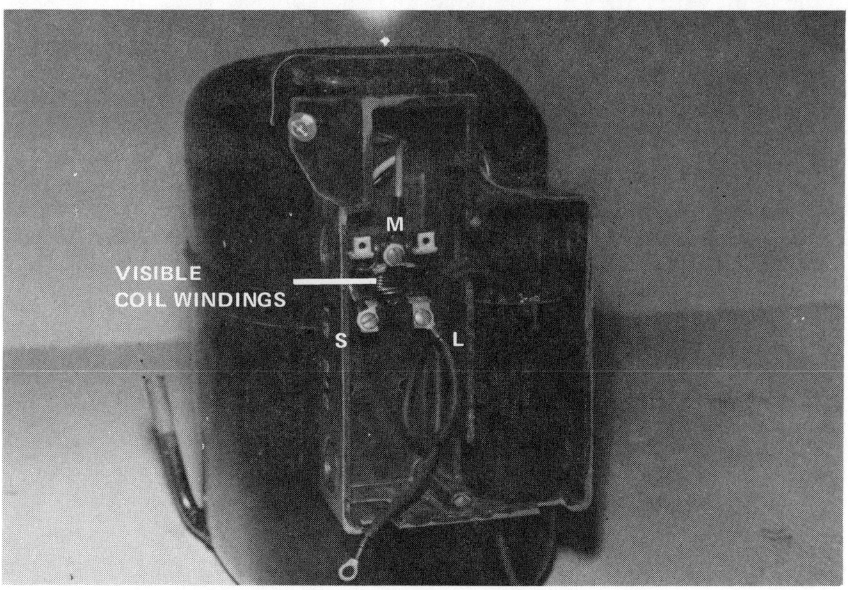

FIG. 6-5. Current type relay is identified by the visible coil and by terminals marked S, L and M.

not sufficient to hold the switch contacts of the relay closed: the switch opens and the start winding is de-energized. All of this occurs in something less than one second.

TROUBLESHOOTING LOW STARTING TORQUE COMPRESSORS

There are four sources of trouble with low torque compressors: wire failure, overload failure, compressor failure, and relay failure.

The lead wires are generally stranded, plastic-insulated wire with push-on or eyelet type terminations. Problems can arise if the insulation becomes frayed, or dried out and cracked, causing a short circuit. Loose wire connections or corrosion buildup at the connections can result in total blockage of current flow or intermittent interference. Intermittent problems are much harder to pinpoint than a totally open circuit. Visual inspection, with an ohmmeter check for continuity, will usually enable you to determine the adequacy of the wiring.

The overload, Figure 6-6, is nothing more than a single pole, single throw switch. An ohmmeter will enable you to see whether or not it is closed. Failure is usually the result of fusing of the contacts. Once the contacts fuse together, the compressor no longer has protection. It's a good idea to change the overload whenever you replace a compressor. Sometimes the arcing that usually causes fusing may burn the contacts to the point where they are no longer making contact. With the overload

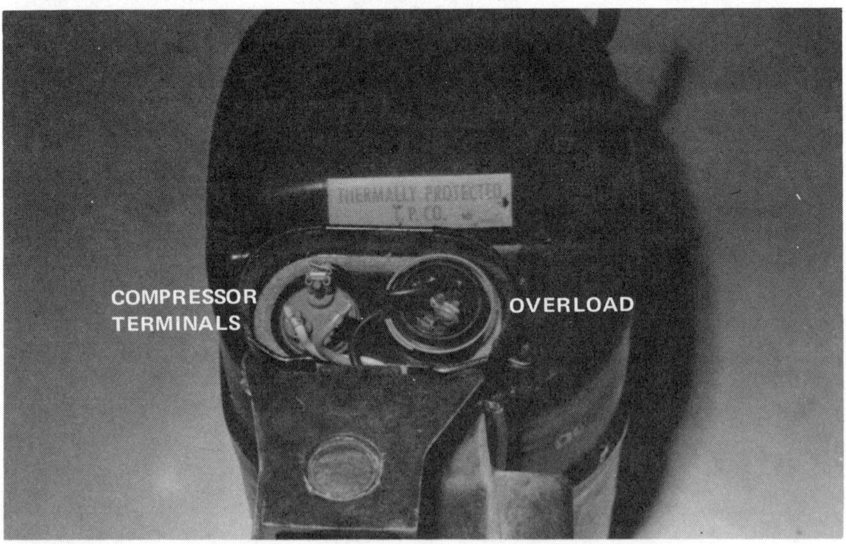

FIG. 6-6. Bi-metal overload switch is connected between the compressor terminal and the common lead of the motor.

open, the compressor will never start. A final problem is a loss of calibration. If the bimetal disc that senses current and temperature opens when the compressor is not unduly hot nor drawing excessive amperage, it has lost its calibration and must be replaced. An ammeter reading of compressor current is useful in detecting this problem. Make sure that the system superheat is not excessive, so that the motor will be properly çooled. If the current matches the nameplate and superheat is 10° ot 15° and the overload trips, the problem must be in the overload itself.

The compressor may have other mechanical or electrical problems. In many cases, a final determination is based on eliminating some easily checked possibilities and then assuming that the cause of the problem must be one that cannot be readily checked. The ohmmeter is a valuable tool in checking the compressor. Check the resistance of the start and run windings, Figure 6-7. An infinite resistance indicates an open winding; a zero resistance indicates a shorted winding; in either case the compressor must obviously be replaced. An ohmmeter reading in between these extremes poses a problem. If you have a table of the manufacturers'

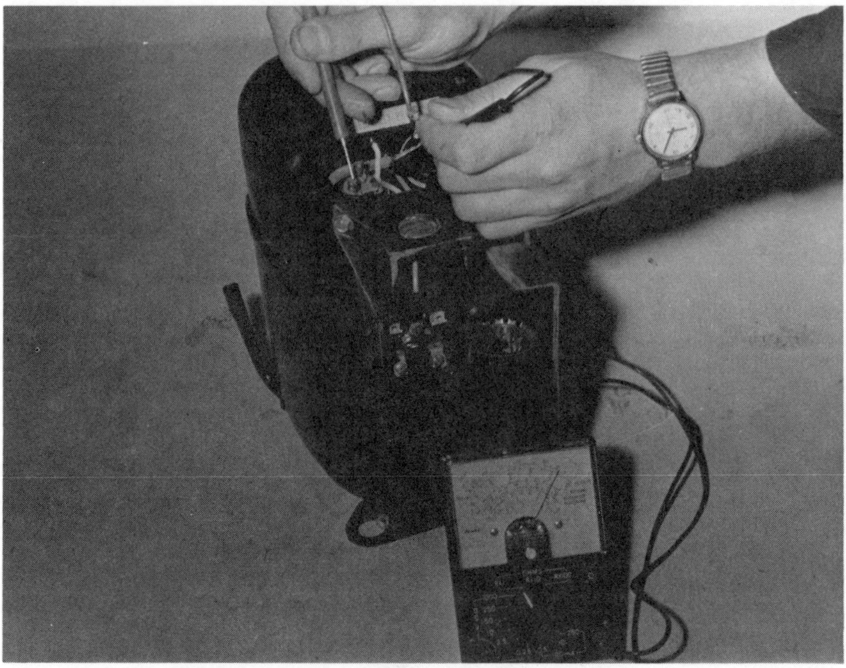

FIG. 6-7. Checking continuity of the motor windings with a low scale ohmmeter. It is always good practice to disconnect all wires from the motor terminals before testing.

values for winding resistance, you can compare actual and stated values. Any variation from stated value will usually result in erratic performance. If you do not have the manufacturers' values, there is not much you can positively conclude. If the problem is a minor short however, you can eventually, by deductive reasoning, still determine that the compressor is at fault.

The ohmmeter should also be used to check for a grounded compressor. Grounded means that a path exists for the current to flow from the compressor windings, through copper tubing, to ground. One ohmmeter lead on a compressor terminal and the other on the suction line will give you an indication. Any resistance value, other than infinite, will be a cause for replacement.

If all components check out well and the compressor still doesn't run properly for any length of time, use a snap-on ammeter, Figure 6-8, and check the running current. If the current is consistently high, even though compressor cooling is adequate, voltage is within 10% of nameplate rating, and head and suction pressures are normal, then the compressor must be replaced. You may never know what the real problem was, but

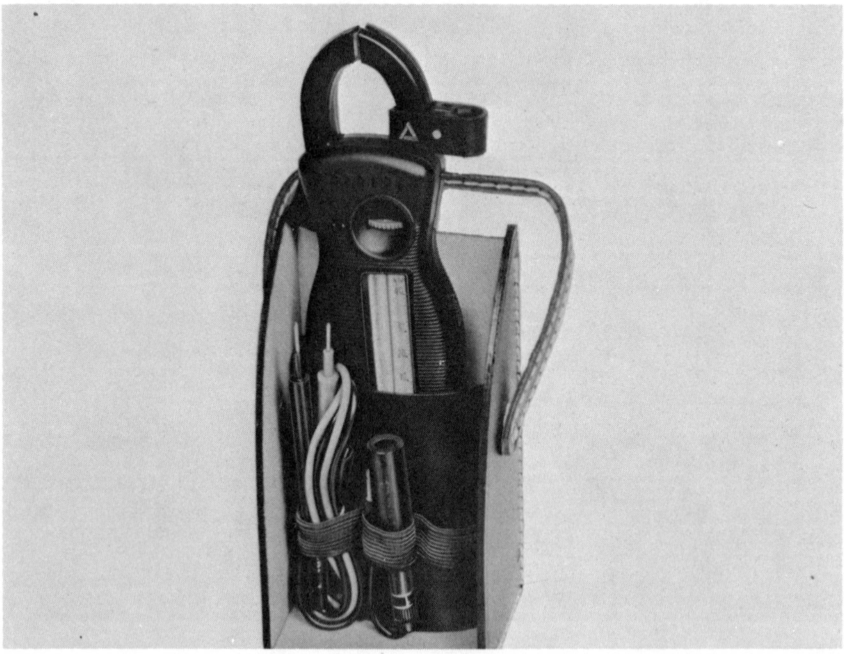

FIG. 6-8. Snap-on ammeter can also be used to check voltage and resistance. *Courtesy Amprobe Instrument.*

at least your customer will have his refrigeration back.

The current relay consists of a coil and a set of switch contacts. Two typical problems with this relay are that the contacts either don't pull in on start-up or they don't drop out when the compressor has come up to speed. By disconnecting all the wires from the relay and checking continuity with your ohmmeter across L and S you can many times solve the problem. Continuity usually will indicate the contacts are welded together and therefore will never drop out. No continuity doesn't tell you much, except that the contacts aren't welded. By putting the leads of a voltmeter across C-S on the compressor and energizing the compressor, you can tell whether or not the start winding is being energized. A voltage reading on start-up indicates the relay is pulling in, no reading indicates it is not pulling in and should be replaced.

A quick check to see if the relay is doing its job is as follows:

1) Disconnect lead from S on relay.
2) Using insulated pliers, hold lead S onto L connection of relay.
3) Energize unit.
4) After one second, pull S lead away from L terminal of relay. CAUTION: Make sure you don't ground out the S lead against any metal or other wires.

In effect, this manual operation is replacing the action of the relay. If the compressor runs smoothly after you've done this, then the problem is bad relay; if the compressor still doesn't start, you'll have to look elsewhere for the problem.

Knowing how each of these components interact should enable you to accurately determine problems in the electrical circuity of reach-in refrigerators using low starting torque compressors. You will probably find that, once the problem is determined to be in the compressor circuity, the order of probability of failure is: (1) relay (2) compressor (3) overload (4) wiring.

One last word. If all components look in good shape and the compressor will still not start, you might try giving the compressor a little starting jolt. Disconnect the S lead from the relay and install a start capacitor of about 100 mfd, 330 volt rating between the relay and the start winding. Attempt to start the compressor several times. If you can get it started, remove the capacitor and see if it will start and run on its own. Many times this technique has proved useful in squeezing a bit more life out of a weakening compressor. If it will only start with the capacitor in the circuit, or if it doesn't start under any condition, it's time for a replacement. This technique is particularly useful with compressors that have been idle for an extended time and just need a little more torque to get started and get the lubrication flowing.

HIGH STARTING TORQUE COMPRESSORS

Figure 6-9 shows a high starting torque compressor typically found with TXV refrigeration systems. A noticeable difference between this

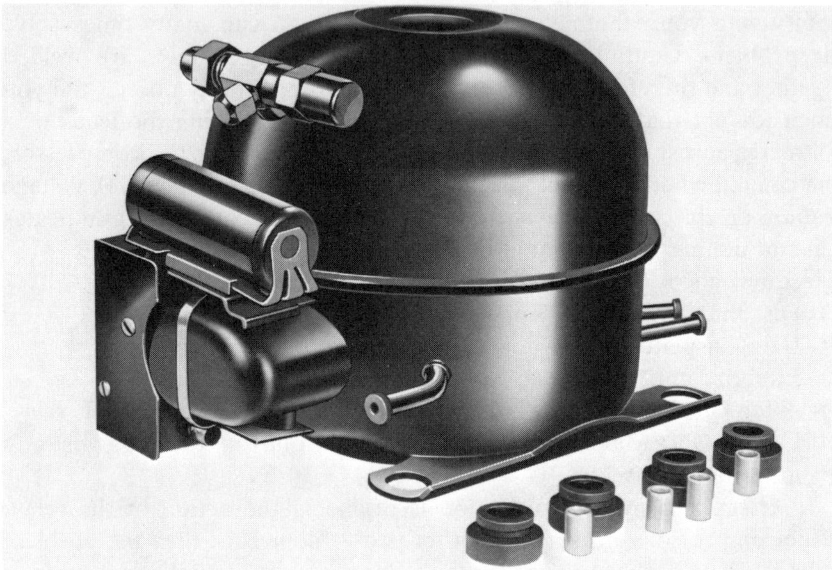

FIG. 6-9. Starting capacitor is shown mounted over the terminal cover of a high torque compressor. *Courtesy Tecumseh Products Co.*

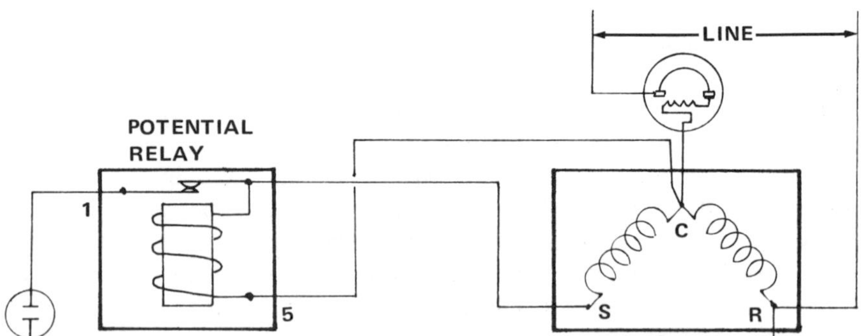

FIG. 6-10. Wiring diagram showing a start capacitor and potential relay used with a high torque compressor.

compressor and the low torque compressor is the addition of a start capacitor above the terminal box. The starting relay on this particular compressor is also of the current type. However, potential relays are used almost exclusively on larger compressors.

The previous discussion of wires, overload and compressor is fully applicable to the high starting torque compressors. A discussion of the potential relay and starting capacitors is now in order. Figure 6-10 shows the schematic wiring diagram for the high torque compressor using these components.

The potential relay has a coil and a set of switch contacts not too unlike the current relay. In the de-energized position, the switch contacts across terminals 1-2 are closed. When the power is turned on, both the start and run windings are energized. As the compressor comes up to speed, an induced voltage, somewhat greater than line voltage, is generated by the start winding. This voltage is imposed on the relay coil at terminals 2-5. When the compressor has come up to approximately 90% of full speed, the voltage is sufficiently high to energize the relay coil, pulling open the switch contacts 1-2. This, in turn, de-energizes the start winding. As long as the compressor is running, the induced voltage across C-S is present and the potential relay contacts will remain open.

The function of the start capacitor is to provide starting direction to the compressor motor by shifting the current phase in the start winding to simulate two-phase operation. In low torque compressors, a similar effect is obtained by using split phase motors. However, the torque of that type motor is quite low, compared to the capacitor start type.

TROUBLESHOOTING HIGH STARTING TORQUE COMPRESSORS

The start capacitor is designed to be in the circuit for no more than three seconds on start-up. It is designed for no more than 20 of these three-second starts per hour, for operation in an ambient of less than 130°F, and should not be subjected to voltages greater than 125% of design. If it always works within these limits, it should have a service life of over 100,000 hours. In the real world, problems such as rapid cycling off and on—due to low refrigerant charge, a defective relay, voltage fluctuations or possibly high ambient operations, particularly in summer—make capacitors a rather common replacement item. Generally a bad start capacitor can be visually identified by oil leaking from its casing. Most, Figure 6-11, have a rubber relief plug that will bulge or blow out completely, giving a good tell-tale indication.

Electrically, the capacitor can be checked by using an ohmmeter on the *100,000 ohm scale*. The needle will rapidly deflect toward zero ohms, then slowly climb toward infinite resistance, if the capacitor is good.

FIG. 6-11. Electrically checking a start capacitor. If the rubber plug in the center of the capacitor is bulged or leaking oil, the capacitor should be replaced.

Remember, use the 100,000 ohm scale or the readings you get may be misleading.

The potential relay, Figure 6-12, has terminal markings 1, 2, 4, 5, 6, with 4 and 6 being dummy terminals that are connected to nothing within the relay. They are provided for convenience in wiring other components in the complete system. Terminals 1, 2 and 5 are wired as shown in the diagram and, regardless of manufacturer, this notation is followed.

As with the current relay, the two types of failures you will find are that the contacts are open, keeping the start winding from being energized, or that they are welded shut, keeping the winding from dropping

FIG. 6-12. Potential relay with an all-purpose mounting bracket. Note the schematic stenciled on the case.

out when the compressor comes up to speed. Occasionally, the coil between 2-5 may also be burned out, but this is rather rare.

A simple continuity check across 1-2 will tell whether the contacts are opening or not. When the unit is initially energized, there should be no voltage reading across 1-2. When the compressor comes up to speed and the switch opens, a voltage reading should be obtained. A resistance reading across 2-5 with an ohmmeter should give a reading between 1,000 and 15,000 ohms. A zero or infinite reading shows a defective relay.

There are many manufacturers's indentifications for potential relays, yet a fairly small number of general purpose replacements can be carried to meet just about any need. The trick is to determine which replacement will do the job. Once you have determined that the starting relay must be replaced, you can proceed as follows:

(1) Remove the lead going to terminal 1 of the relay.
(2) With a pair of *insulated* pliers hold the lead against terminal 2.
(3) Energize the unit and, within one second, pull the wire away from terminal 2. CAUTION: Don't touch this wire against any metal as you will short it out.
(4) Measure the voltage across 2-5, which will be something greater than line voltage.

FIG. 6-13. This 6-pack of relays replaces a wide selection of manufacturers' models. *Courtesy Motors & Armatures, Inc.*

Potential relays have several characteristics which are usually marked on the box. The coil of the relay has a continuous voltage rating. You must select a replacement relay with coil rated at something more than the voltage you read across 1-2. If not, the replacement relay you select will burn out rather quickly. The pick-up voltage of the relay is the voltage at which the switch will open, de-energizing the start winding. A pick-up voltage of not less than 10% below the 1-2 reading you obtained should be selected. The drop-out voltage is the voltage at which the switch closes when the compressor is de-energized and comes to a halt. A value less than half of the 1-2 reading should be selected. Having a fairly limited selection of replacement relays will enable you to replace a wide selection of manufacturers' models, Figure 6-13.

FAN MOTORS

Other than compressors, the only motors you will run into on reach-in refrigerators are evaporator and condenser fan motors, Figure 6-14. They come in an assortment of shapes and sizes. Typically, they cannot be lubricated and require little maintenance service. They start when the box is plugged in and run until they die, generally quite a few years later. About the only maintenance that can be performed is to keep them clean and free of any dirt that might inhibit the flow of cooling air over them. This is particularly true of condenser fan motors that may be in dirty locations.

FIG. 6-14. These small motors are widely used with evaporator and condenser fans.

Many times a motor was thought to be burned out when all that had happened was that the bearing lubrication had dried out and the motor, having low torque characteristics, would not turn. A few drops of oil, and some trial spins by hand, got the motor free and spinning again.

The nameplate data, Figure 6-15, is vital in selecting a replacement

FIG. 6-15. Typical motor nameplate data is stenciled on the frame.

motor and every effort must be made to match operating characteristics. The voltage and rpm should be exactly matched. Horsepower or full load current may both be shown, or only one may be shown as an indication of the power of the motor. Either the same or a higher power motor should be selected. Choosing the next smaller size, if an exact replacement isn't available, may very well lead to a repeat failure.

Physical dimensions are important. Nothing is as frustrating as having a perfect motor in your hands, ready for replacement, and not having it fit. There is no dependable standarization of mounting dimensions on these small motors, so it is best to take the failed motor with you when searching for a replacement or at least having the key mounting dimensions written down.

One final word pertains to motor rotation. Strangely enough, clockwise (CW) and counter-clockwise (CCW) notations mean different things to different manufacturers. Some arrive at clockwise rotation by looking at the shaft end of the motor and others look at the end opposite the shaft.

Make sure you know how the motor should be rotating and make sure, when ordering a replacement, that you and your order-taker are communicating effectively.

Incidentally, if you do get a wrong-rotation motor, you can't leave it in and walk away, as air flow will be tremendously reduced. Nor can you turn the fan blade around. The blade was made for a particular rotation and there is no way to modify it. If you have a properly designed blade for the new rotation, you might get away with changing blades. Even so, you have to evaluate the new air flow patterns you've created to see if they will do the job.

7

Reach-in Freezers

There are several differences between reach-in freezers and refrigerators that are not obvious to the casual observer. Figure 7-1 shows a reach-in freezer. In external appearance it is identical to a refrigerator, Figure 1-4. Since freezers generally are required to hold temperatures between $-5°F$ and $+5°F$, we would expect evaporator temperatures around $-20°F$ to $-10°F$. Since most freezers of this type use R-12 refrigerant, the low side pressure will be around 0 to 5 psig.

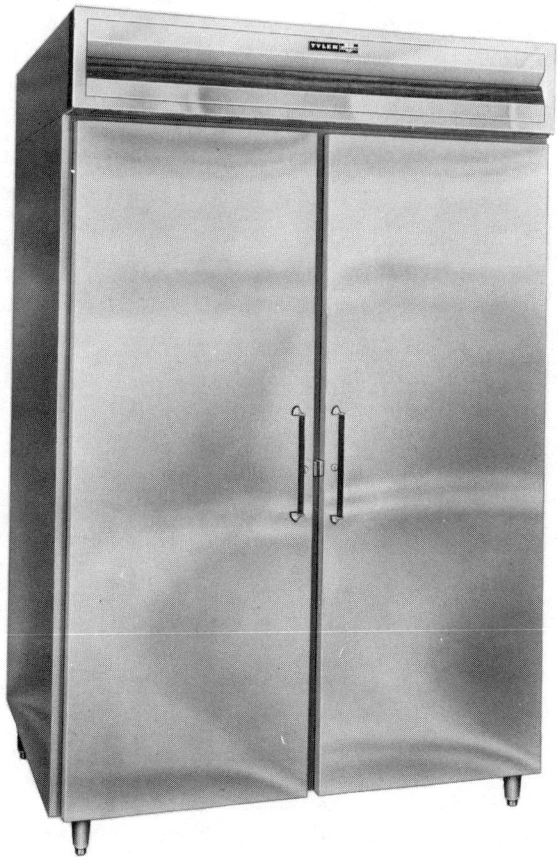

FIG. 7-1. Two-door, reach-in freezer. *Courtesy Tyler Refrigeration Corp.*

Since we're experts in the use of the Mollier diagram by now, we'll take a look at this freezing cycle superimposed on the refrigeration cycle developed earlier, Figure 7-2. You will note several differences in a typical freezing cycle compared to a refrigeration cycle:

1. Flash gas is substantially greater.
2. Net refrigerating effect is lower.
3. Heat of compression, hence power requirement, is greater.
4. Discharge temperature is higher.

Summarizing: freezing is a less efficient, more costly, procedure than refrigerating.

Since food that is properly frozen and held at low temperatures can be stored for quite long periods of time, it is safe to say that, despite their relative inefficiency, freezers are here to stay. In fact, the abundance of packaged frozen goods, and the economy of bulk buying and long-term storage, make the economy of a freezer a real one, despite possible increased operating cost. One restaurateur, famous for his apple pies, actually makes his winter requirement of apple pies in October, using freshly picked apples, and freezes them, by the hundreds, until needed.

Both TXV and capillary tube condensing units are used. The major

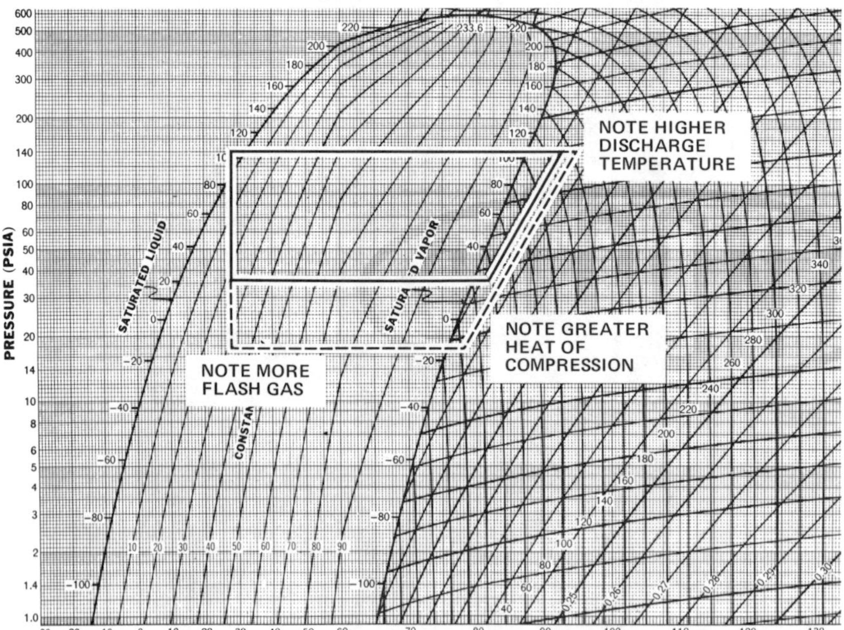

FIG. 7-2. Mollier diagram comparing a low temperature refrigeration system with a medium temperature system.

difference in these condensing units is the compressor, which is designed for low temperature operation. Subjecting these compressors to the suction pressures typical of normal refrigeration (20-25 psig) will cause them to continually draw high current, leading to eventual cycling on overload and possible failure. While some compressors may be able to do both jobs, most cannot. The compressors are similar in appearance however, and all remarks previously made concerning wiring, relays, overloads, etc. are perfectly valid for both refrigeration and freezing.

TYPES OF REFRIGERATION

Up to this point, the term refrigeration has been used in two ways. Refrigeration is the process of taking heat from where it is not wanted and moving it to where it is not objectionable. Refrigeration is also maintaining storage temperatures of about 35°F, compared to freezing, where 0°F temperatures are usual.

From this point on, various types of refrigeration will be classified as: low temperature, with storage temperatures around 0°F; medium temperature, with storage temperatures around 35°F; and high temperature, with air temperatures around 50° to 60°F, more specifically, air conditioning. Low temperature applications are, therefore, freezers. Medium temperature (sometimes also referred to as commercial temperature) applications are refrigerators.

Industry-wide standarization in the air conditioning field has seen the capacity of high temperature systems specified in Btu's per hour or tons (1 ton = 12,000 Btuh) which is a fairly precise way of describing the cooling capacity of the equipment. No such standardization exists for low and medium temperature systems. Although a trend is afoot to use Btuh universally, old habits die hard, and other descriptive terminology still persists. Perhaps the most popular of these old terms is the *horsepower.*

In high temperature refrigeration, one horsepower equals approximately one ton or 12,000 Btuh. A one horsepower compressor running a medium temperature system will only produce about 8,000 Btuh of cooling and that same one horsepower compressor on a low temperature system will produce only about 4,000 Btuh of cooling. This makes sense if you look back at the Mollier diagram, Figure 7-2. Note that, as we move from a medium to a low temperature operation, the refrigeration effect is reduced and power consumption increased due to greater heat of compression.

One manufacturer's catalogue rates a 1/3 horsepower condensing unit, typical of that used on a reach-in refrigerator, at 2,540 Btuh with an evaporator suction temperature of +20°F. A different 1/3 horsepower unit, used on a freezer, has a capacity of 1,380 Btuh at a suction temperature of −20°F. It is therefore important to keep the application in mind when selecting a replacement condensing unit, or equally important, when selecting a condensing unit to match a remote reach-in.

DEFROSTING

Since the air temperature in a low temperature box is below freezing, we cannot count on the air to melt accumulated ice from the coil. A positive means of evaporator defrosting must be provided.

While there are still freezers made with natural convection evaporators, the bulk of the units in the field today have blower coils. The natural convection units are usually defrosted by emptying the box, leaving it off for a while, and scraping the ice off with a *non-metallic* scraper. Since this requires effort that many equipment owners would prefer not to expend, the requirement for automatic defrosting exists.

Defrosting cycles can be examined by considering them to be composed of three parts: Initiation, Method, Termination.

Initiation

The most widely used method of initiating the defrost cycle is by

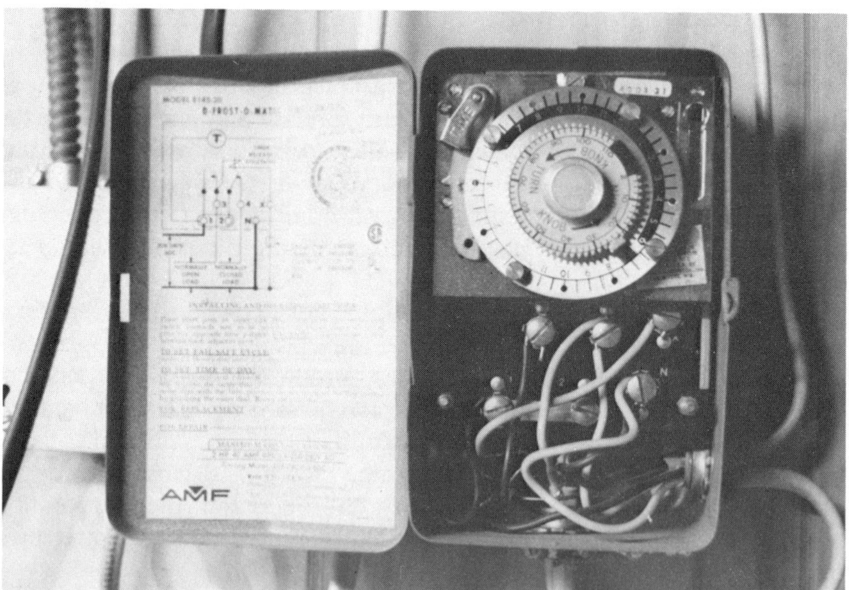

FIG. 7-3. Defrost time clock is used with both reach-in and walk-in freezers.

FIG. 7-4. This non-adjustable defrost timer is used with domestic freezers and smaller commercial freezers. The defrost cycle is factory set.

means of a timer, Figure 7-3. This timer is usually adjustable and any number of cycles can be set for a 24-hour period. A good number of boxes use the type of timer shown in Figure 7-4 which cannot be adjusted. It is factory set for a specified number of cycles per 24 hours.

Occasionally, a more exotic initiation technique may be encountered, such as a device that counts the number of times the freezer door is opened. After a specified number of openings, the cycle is initiated. This is not as bizarre as it may sound since buildup of ice is directly related to opening and closing the door. Another device senses the pressure differential across the coil. As the differential increase, due to buildup of ice, a pressure switch trips and initiates the cycle.

Method

Two techniques account for practically all defrost systems. A schematic of the hot gas defrost method is shown in Figure 7-5. Upon initiation of the cycle, a solenoid valve connecting the hot gas line to the evaporator coil is energized, Figure 7-6. The hot gas takes the path of least resistance and melts the ice as it flows through the evaporator.

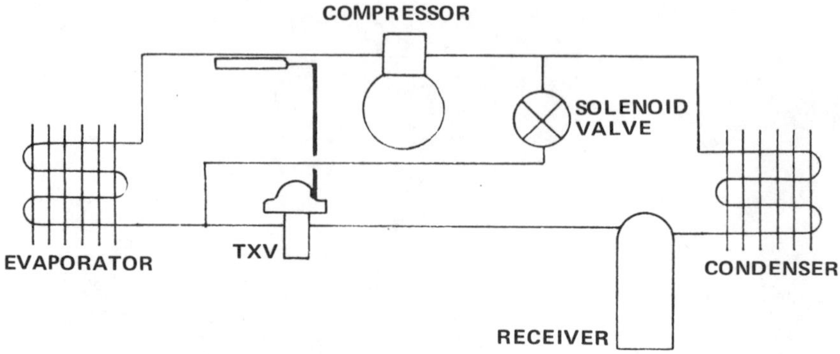

FIG. 7-5. In the hot gas defrost system, vapor from the compressor discharge is shunted to the evaporator through a solenoid valve.

FIG. 7-6. Solenoid in the hot gas bypass line.

While this works effectively in properly designed and installed units, several shortcomings have resulted in the hot gas defrost system falling into disuse in small commercial boxes. High suction pressures, for an extended period of time, can overload the compressor. Installation in a low ambient location can result in low discharge temperatures and excessive defrost time or incomplete defrosting. Hot gas in the evaporator can condense to a liquid and flow through the suction line to the compressor, causing damage.

The most popular technique is the electric defrost system. The timer de-energizes the compressor and the fan motor and energizes electric heaters mounted in the evaporator coil. The heaters melt any ice accumulation.

Termination

The simplest way to terminate the cycle is to have the time clock set for a specific duration, after which the cycle is terminated. Ideally, the cycle lasts only long enough to melt all the ice. If the cycle is too short, ice remains and efficiency is impaired. If the cycle is too long, excessive heat is generated, possibly affecting food quality and certainly adding to the load on the refrigeration system when normal cooling is resumed. Since a timer can sense neither of these conditions, some other means is used to terminate the cycle. In many arrangements, the timer is used as a backup or fail-safe terminator.

A popular termination device is an integral temperature sensor mounted on the surface of the coil. It may be set to terminate the cycle when the coil temperature reaches approximately 40°F. This temperature will not be reached until all the ice is melted.

Pressure termination is also built into some time clocks. The pressure in the evaporator will build rapidly once all the ice on it has been melted. A tube from the time clock is connected to the suction side of the compressor and will sense this rapid increase in pressure. At a predetermined setting, the sensor will cause the switching mechanism within the clock to terminate the defrost cycle and initiate the freezing cycle.

The timer and temperature terminator can be used with both electric and hot gas defrost, while the pressure terminator can only be used with electric defrost.

DEFROST CONTROL WIRING

Figure 7-7 is a typical wiring diagram for an electrically defrosted reach-in freezer. The switches are shown in the freezing cycle, assuming that the Fused Disconnect Switch is closed.

The Clock Motor timer initiates the defrost cycle by opening switch A and closing switch B, both shown near the Switching Solenoid.

Opening switch A interrupts the circuits to the Compressor Motor and the Fan Motor.

Closing switch B completes a circuit through terminal 3, to the Heater and back through the Heater Safety Control, which is typical safeguard with all electric heaters.

The defrost cycle will continue until the clock switches from circuit B to circuit A or sooner, if the Defrost Terminator & Fan Delay Control

TL WIRING DIAGRAM – 115 VOLTS

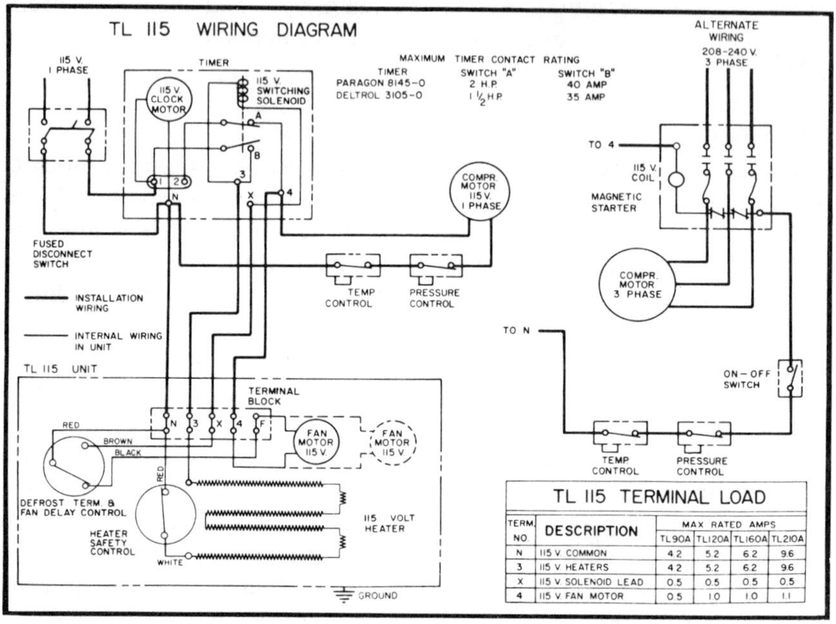

TL 115 TERMINAL LOAD

TERM. NO.	DESCRIPTION	MAX RATED AMPS			
		TL90A	TL120A	TL160A	TL210A
N	115 V. COMMON	4.2	5.2	6.2	9.6
3	115 V. HEATERS	4.2	5.2	6.2	9.6
X	115 V. SOLENOID LEAD	0.5	0.5	0.5	0.5
4	115 V. FAN MOTOR	0.5	1.0	1.0	1.1

FIG. 7-7. Typical wiring diagram for an electric defrost system. *Courtesy Bohn Heat Transfer Div., Gulf + Western Mfg. Co.*

senses that the coil temperature has reached a point where all the ice on the coil has melted. At that point, the Defrost Terminator will thermostatically switch to close a circuit between the Red and Brown terminals and complete a circuit to the Switching Solenoid.

When energized, the Switching Solenoid will open circuit B and close circuit A. In so doing, it opens its path to neutral, through the heater circuit, and is de-energized.

Whether the defrost cycle is terminated by the Clock or the Solenoid, only the compressor will be energized when circuit A is restored, since the circuit between the Red and Black terminals is still open.

When the coil becomes chilled, the Fan Delay Control will switch, again thermostatically, to close the fan circuit between terminals Red and Black. When the fan starts, it may very well move enough warm air over the coil to reopen the Fan Delay Control. In fact, the Fan Delay may open and close several times before the coil is at normal operating temperature and fan operation is continuous.

A very similar wiring configuration can be drawn for a hot gas defrost system. They differ in that a hot gas solenoid valve would replace the

Heater and the Compressor Motor will continue to run during the defrost cycle.

CONDENSATE DISPOSAL

The defrost cycle melts the ice from the coil and produces condensate that must be then disposed of. If the box is close to a floor drain or waste line, a logical solution is to run a drain from the coil to the building drain. This is, in fact, done in many installations.

The design of any condensate disposal system should assure that the drain line within the box is not long nor does it contain a trap. The drain line is passing through 0°F air and the water flowing through it could possibly freeze.

Usually, the coil drain pan and piping are warmed by strategic placement of the electric heater, or the hot gas defrost line, to assure positive runoff during the defrost cycle. Any freeze-up or obstruction will allow condensate to back up into the drain pan and lead to a massive ice buildup in the pan and on the coil.

When a building drain is not convenient, two other means of condensate disposal are available.

A condensate disposal pump, Figure 7-8, can catch the runoff from the coil and pump it to a building drain, utility sink, outdoors, roof, etc.

FIG. 7-8. Typical condensate disposal pump. *Courtesy Watsco, Inc.*

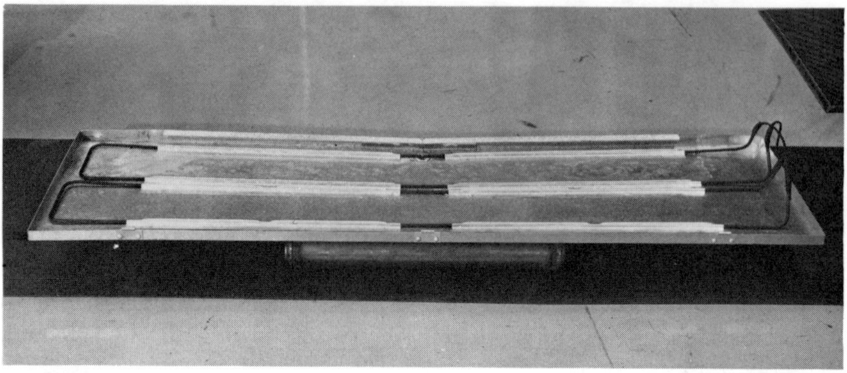

FIG. 7-9. Two electric heaters mounted in the drain pan of an evaporator blower. *Courtesy Bohn Heat Transfer Div., Gulf + Western Mfg. Co.*

Condensate pumps are available in a number of sizes to accomodate any number of refrigeration units.

Evaporative pans are used to store condensate until it evaporates. To hasten evaporation, heat is provided by any number of sources. Some units have the compressor discharge line routed along the bottom of the pan, to provide heat and to enjoy some slight theoretical increase in system efficiency. Electric heaters, Figure 7-9, are available in some pans. In other applications, warm condenser air is directed across the surface of the condensate lying in the pan to increase the evaporation rate.

Problems with condensate disposal systems are usually indicated by water on the floor around the box or when ice has built up around the base of the evaporator coil. Typically, a blockage of the drain line is caused by dirt or by a stray particle of food; or the condensate pump may be blocked, its motor inoperative, or its float switch defective. With evaporative systems, the source of heat, namely the electric heater, has become inoperative.

EVAPORATORS

While condensing units in low and medium temperature refrigeration are quite similar, there are several differences in the evaporator coils.

The major difference is the addition of electric heaters. Externally, there is little apparent difference in the appearance of an evaporator coil used in low temperature applications. However, one look at the coil surface will reveal that heaters are either surface mounted or imbedded in the coil core. There may be an additional heater in the drain pan, Figure 7-9, to insure free flow of condensate.

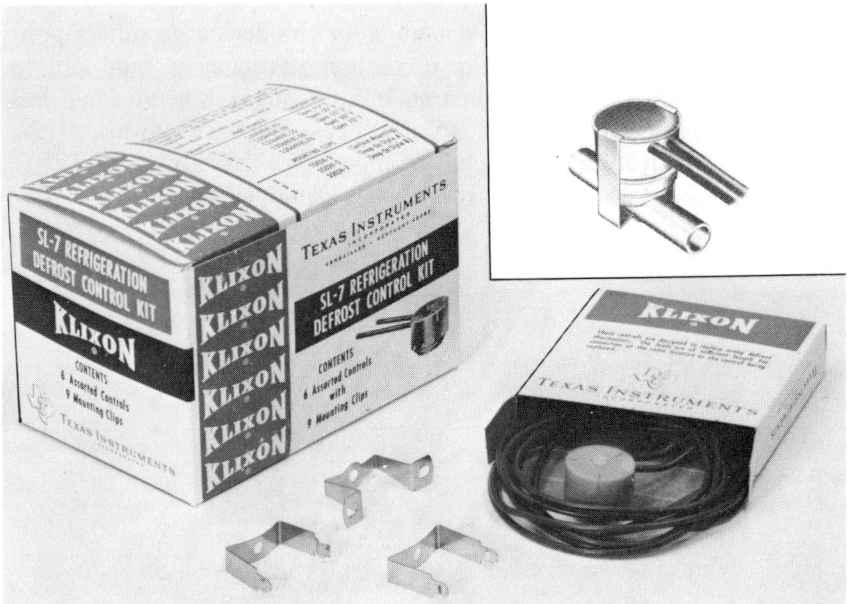

FIG. 7-10. Defrost termination thermostat with a variety of mounting clips. *Courtesy Motors & Armatures, Inc.*

The coil used in hot gas bypass systems will be quite similar to a medium temperature coil, the main difference being an additional inlet on the downstream side of the refrigerant flow control to permit hot gas to flow to the coil, bypassing the control. The path of this hot gas may also flow through the drain pan, again to permit free flow of condensate.

Electrically, the evaporator coils are similar, in that one or more motors blow air across the coil. They are dissimilar to their medium temperature cousins, in that a temperature switch, Figure 7-10, may be mounted on the coil to serve two functions. This switch prevents the fan from running until the coil temperature has dropped low enough to provide cooling. Air blowing across a coil that has just been heated, to melt off accumulated frost, could possibly melt part of the contents of the box. The time delay, provided by this temperature switch, insures that once the fan comes on, usually several minutes after the compressor has been running, only cool air will be blown over the box contents.

The second function of this switch is to terminate the defrost cycle by signalling the timer when the coil is sufficiently warm to guarantee that all ice has been melted. Some such switches provide an adjustment to regulate the length of the defrost cycle, others are factory set and not adjustable.

Sometimes both functions are contained in one device. In other applications, two separate switches are used. It is particularly important to remember the fan time-delay function because many a serviceman has pulled out a perfectly good fan motor because he wasn't aware of the existence of the delay and assumed the motor was defective.

Hot gas defrost evaporators are similarly equipped.

8

Walk-in Refrigerators

Reach-in refrigerators, as was mentioned earlier, are used very widely in commercial applications. They are intended for short-term storage of foods and therefore have temperatures on the order of 38 to 45°F. They are convenient to use, everything being within reach when the door is opened, and their size permits them to be located quite close to the food preparation or serving areas. The compressor is usually about 1/3 horsepower and ranges up to as high as one horsepower on six-door boxes. The vast majority of reach-in refrigerators have 115v single-phase condensing units.

Walk-in refrigerators are designed primarily for long-term storage. While they may be entered many times a day in actual practice, this is usually due to poor management of the facility. Long-term refrigerator storage temperatures are usually between 33 and 39°F. These boxes are not nearly as convenient to use as reach-in refrigerators because of their massive doors and because their size usually requires a remote location out of the main flow of kitchen traffic. Small walk-in refrigerators may use one horsepower, 115v single-phase condensing units but, as size increases, 230v single-phase and 208v three-phase units are used.

The components of a walk-in refrigeration system are quite similar to the reach-in box. The capillary is rarely used as the refrigerant flow control, most systems employing the thermostatic expansion valve. Perhaps the major difference is that the TXV employs an external equalizer, in many cases.

Electrically, the walk-in refrigerator is significantly different from the reach-in. Most reach-in's use line voltage controls in series with the compressor. In walk-in's of any appreciable size, we find that the compressor and fan motors are usually energized by contractors or motor starters. The thermostats and pressure controls are usually in control circuits, sometimes 24v, but more often 208v, and their operation energizes the coil of the contractor or starter. We'll spend some time on this aspect of walk-in refrigerators as it is particularly important.

BOX CONFIGURATIONS

Many walk-in boxes in use today are built of wood. Most of the newer boxes, however, are of all-metal construction, Figure 8-1. The lowest priced will be steel construction with a baked enamel finish. The mid-

FIG. 8-1. This large walk-in refrigerator can accommodate a fork lift truck. *Courtesy Elliott-Williams Corp.*

priced range will be a box with unfinished aluminum on the walls and ceiling and a steel floor. The most expensive, and by a wide margin, will be the all stainless steel box. Some experimentation with fiberglass and plastic sheeting is being done, but this construction is not yet commercially important.

In the *good old days* the standard insulation used was 6 inches of cork. There was nothing cheap about cork, so an *innovation* was the use of 6 to 8 inches of fiberglass, similar to that used in residential construction. The newer boxes employ *foamed-in-place* urethane and similar plastic foams of high insulating value. A 2-inch thick urethane wall is equivalent to a 6-inch wall using fiberglass. Due to *foamed-in-place* technology, today's boxes are less massive in appearance and easier to use because of much lighter doors.

Many walk-in boxes were originally designed into the buildings housing them. This is particularly true in schools, hospitals and other permanent facilities. Construction may be concrete block with cork insulation or common bricks with urethane sheet insulation or wood with fiberglass insulation or any other type of construction that comes to mind. The space was designed to be a walk-in refrigerator and nothing else.

This contrasts with the prefabricated walk-in made of panels that are shipped to a job site and then assembled into a refrigerator. While much new construction specifies this type of box, probably many more go into

FIG. 8-2. Packaged walk-in refrigerator with through-wall refrigeration system. *Courtesy Master-Bilt Products.*

locations that originally were designed for something else. A furniture store, converted to a butcher shop, or the pantry of a hospital, modified to provide extra cold storage, are two examples of where a prefabricated box could be used. They are easily erected by two men and will provide long, effective service if not abused.

Two types of refrigeration packages have found wide use in walk-in applications. A preassembled unit, with all components and controls similar in appearance to a through-wall air conditioner, is available. It can mount through a side wall opening, Figure 8-2, or over a ceiling opening. The second type is along more conventional lines. An evaporator coil is hung in the box and is piped to a condensing unit at a remote location. In recent years, packaged condensing units, containing all controls prewired in a master control panel and having a weatherproof enclosure for outside installation, have become available.

REFRIGERATION CYCLE

The full hermetic compressor is used in the vast majority of reach-in refrigerators. Full hermetics find wide use in walk-in boxes but, more

FIG. 8-3. Semi-hermetic compressor may be installed indoors and is available with a weatherproof shroud for rooftop or on-slab installation.

often than not, a semi-hermetic compressor, Figure 8-3, will be used and you will find a surprisingly large number of open type units. The open type unit still enjoys a good reputation among knowledgeable service people in the commercial field, though its popularity is on the wane.

While the equipment itself is larger than any encountered in reach-in boxes, Figure 8-4, the principles of operation are the same. The problems and solutions described in Chapter 1 are fully applicable to the walk-in systems. You might just go back and skim through Chapter 1 now to refresh your memory.

EXTERNALLY EQUALIZED EXPANSION VALVE

The operation of the thermostatic expansion valve depends upon the balancing of three forces, Figure 3-5. The pressure on one side of the diaphragm is directly related to the temperature of the suction line and tends to open the valve. Countering this pressure are two other forces, acting on the opposite side of the diaphragm, that tend to close the valve. One is a spring, preset at the factory, that regulates the amount of superheat. The second force is the pressure of the refrigerant on the evaporator side of the valve seat. While we are actually attempting to control

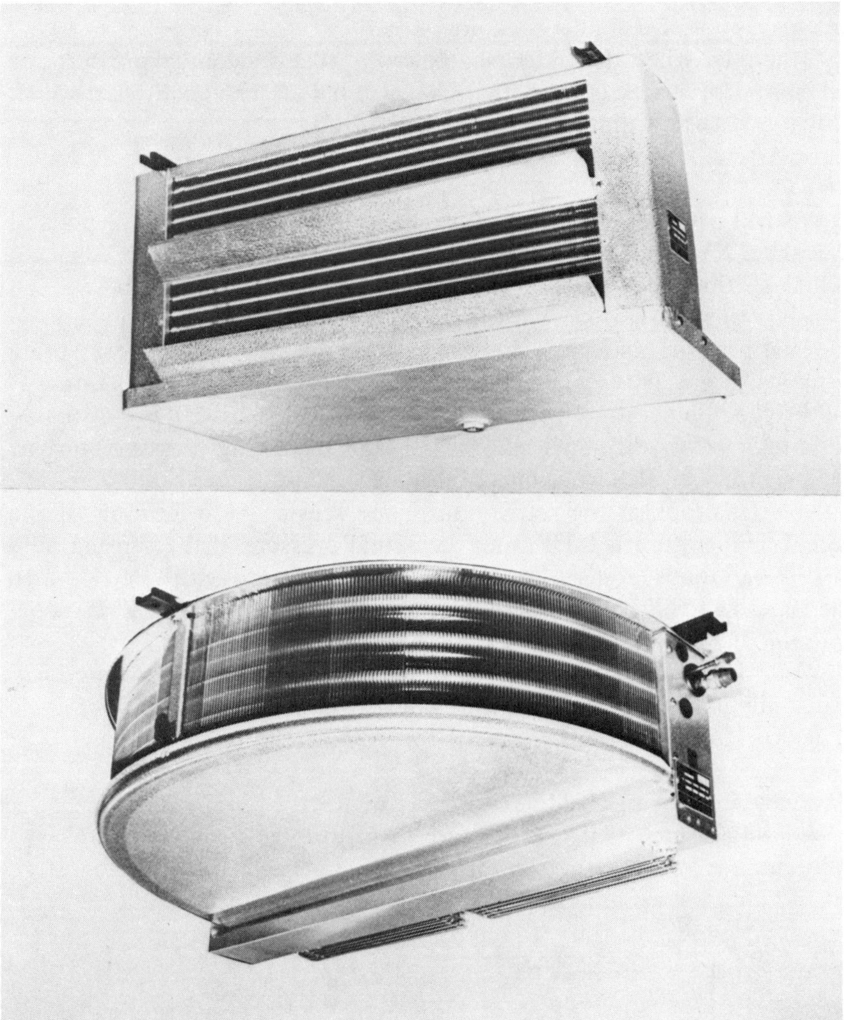

FIG. 8-4. Half-round and standard blower coils are commonly used in walk-in boxes. *Courtesy Climate Control Div., Singer Co.*

superheat at the TXV sensing bulb, located on the suction line *leaving* the evaporator, the pressure acting on the diaphragm is the pressure of the refrigerant about to *enter* the evaporator. On small evaporator coils, typical of those used in reach-in refrigerators, the difference in refrigerant pressure entering and leaving the coil, called pressure drop, is very small. We can get satisfactory control using an *internally equalized* TXV, that is, a valve with the pressure acting on the bottom of the diaphragm being evaporator coil inlet pressure.

Generally, when the difference between inlet and outlet pressure, or pressure drop, exceeds 5 psi, the TXV will behave erratically. Let's look at two cases using an internally equalized TXV.

Case One

We will assume that the pressure drop through the coil is 2 psi and that the TXV is factory set for a superheat of 10°F. If we measure an actual suction line temperature of 30°F at the sensing bulb, we can translate this into a downward force of 28.5 psig. If the actual evaporator inlet pressure under the diaphragm is 21 psi, then the factory spring is set at 7.5 psi. This makes sense because the suction line temperature is 30°F, the temperature corresponding to 21 psig is 20°F (assuming R-12), therefore the superheat is 10°F and the spring pressure must be 7.5 psig for the valve to be in equilibrium.

Now let's look at the refrigerant at the sensing bulb location. If the actual temperature is 30°F, and the actual pressure at the sensing bulb is 19 psig (that's 21 psig less the pressure drop through the coil), then the superheat will be 30°F less the saturated temperature of R-12 at 19 psig, 17°F, or a superheat of 13°F.

Well, that's not bad. Our valve, designed to maintain 10°F superheat, is actually holding 13°F superheat. This is within the typically acceptable range of 10 to 15°F superheat used in commercial refrigeration.

Case Two

We will assume that the pressure drop through a different coil is 8 psig and that we use the same internally equalized TXV set for 10°F

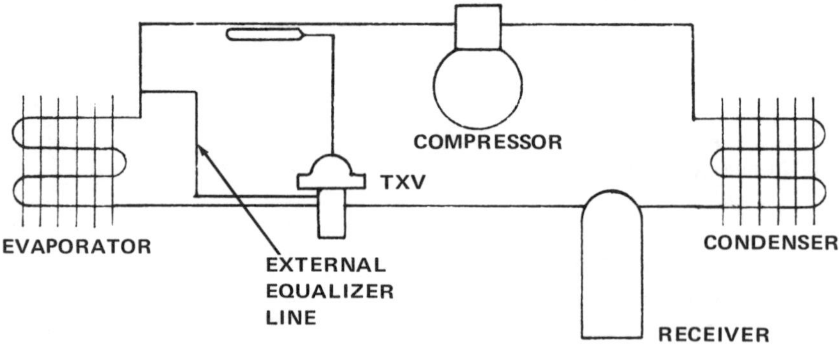

FIG. 8-5. TXV system equipped with an external equalizer.

superheat. If we again assume a 21 psig pressure entering the coil, the pressure leaving the coil will be 13 psig (21 psig less the 8 psig pressure drop). The actual superheat at the sensing bulb will be 30°F less the saturated temperature of R-12 at 13 psig, approximately 7.5°F, or a superheat of 22.5°F.

What has happened is that we have applied a factory set, 10° superheat, internally equalized valve to an evaporator coil with high refrigerant pressure drop and it resulted in a 22.5°F superheat being maintained at the sensing bulb. High superheat, of course, means that not enough refrigerant is getting into the evaporator coil, with a resulting loss of capacity and possibly dangerous decrease in suction gas cooling of the compressor. If the coil has an even higher pressure drop, superheat will increase still further and will result in still worse performance of the system.

Obviously, we need a valve that will maintain 10°F superheat at the sensing bulb when the pressure drop through the coil is high. The answer is the externally equalized valve.

Case Three

Let's review what's been happening. A TXV diaphragm has been acted upon by three forces and achieved equilibrium at a superheat of 10°F. While the valve was in equilibrium in Case Two, and really, sincerely felt it was maintaining a superheat of 10°F, it was in fact maintaining a superheat of 22.5°F, at the suction line close to the evaporator. This is undesirable, as we've seen.

Let's look again at our valve diaphragm and, instead of having the entering evaporator pressure acting on the bottom of the diaphragm, let's run a sensing line, a ¼ inch copper tube, to the suction line near the sensing bulb, Figure 8-5. We'll block off the inlet pressure, so it can't reach the diaphragm, and permit only the suction pressure, after the evaporator, to push on the diaphragm in addition to the spring, Figure 8-6.

We now see 28.5 psig acting downward, a 7.5 psig spring force, and a 13 psig suction pressure acting upward. Since 28.5 psig is greater than 20.5 psig (7.5 + 13) the valve will open, allowing more refrigerant to enter the coil. This increased refrigerant flow causes the evaporator pressure to increase to 29 psig and 31°F saturated temperature. As a result, the temperature at the sensing bulb will also increase. The valve will assume a new equilibrium position when the force acting upward on the diaphragm reaches 28.5 psig. Since the spring force is 7.5 psig, the pressure at the evaporator outlet will increase to 21 psig when equilibrium is reached. This 21 psig pressure corresponds to a saturated refrigerant

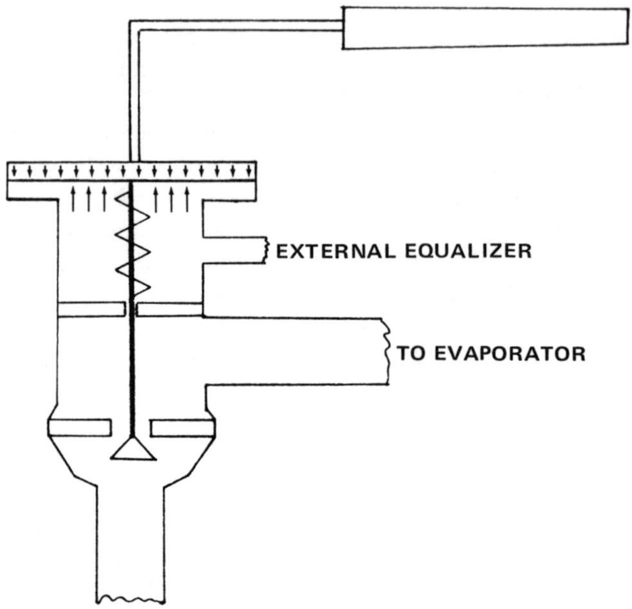

FIG. 8-6. TXV controlled by suction pressure, through the equalizer line, rather than evaporator inlet pressure.

temperature of 20°F. Since the sensing bulb feels a temperature of 30°F, the actual superheat is 10° (30° − 20° = 10°).

This is more like it. Our TXV can control in an acceptable fashion, since the pressure under the diaphragm is close to the actual pressure near the sensing bulb. The use of an externally equalized TXV on an evaporator with a pressure drop exceeding 5 psi results in better superheat control, more effective use of the evaporator surface, increased compressor operating efficiency, and adequate suction gas cooling at the compressor.

Summary

The manufacturer of an evaporator knows the pressure drop through his coil in a typical application. If the pressure drop is excessive, he will provide a connection in the suction line for an external equalizer line. It wouldn't be there if it wasn't needed, so to ignore it and use an internally equalized valve, is to invite trouble. When replacing a TXV, make sure

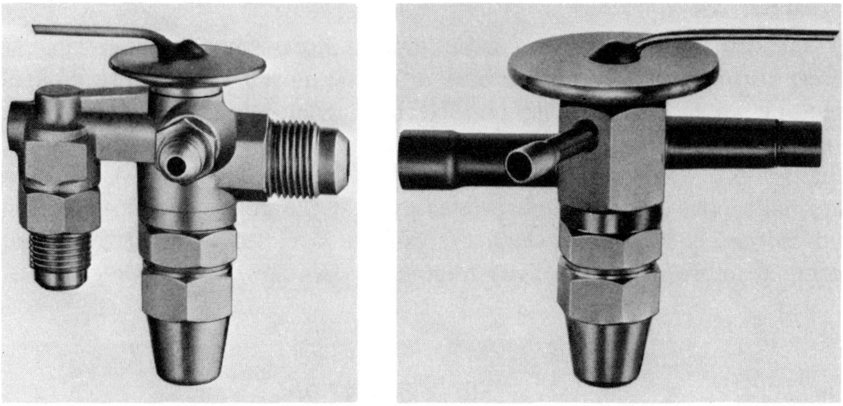

FIG. 8-7. Externally equalized thermostatic expansion valves are easily recognized by the presence of a third port which is a ¼ in. flare or solder fitting. *Courtesy Controls Div., Singer Co.*

you recognize whether or not it is externally equalized, Figure 8-7, and use the appropriate replacement. One final word: an externally equalized valve must be used with an external equalizer line. The fitting on the valve, to which the line is to be connected, *cannot* be capped off. The valve will not work at all if you do.

This was kind of a long-winded explanation of the difference between internally and externally equalized thermostatic expansion valves but, if you followed it through, you should have an insight into how these valves work. I guarantee you will have a better understanding of these valves than probably 95% of the guys in the field today.

ELECTRICAL CONSIDERATIONS

The major difference between walk-in and reach-in refrigerators is in how the compressor is started. The reach-in refrigerator compressor is usually connected in series with a pressure control or thermostat. The control switch will open and close in response to pressure or temperature and break or make the line voltage circuit to the compressor and condenser fans. This is fine when the current draw of the equipment is below 12 to 15 amps. As equipment size increases, however, the electrical arcing at the switches, and the large size of the switches themselves, call for the use of *control circuits*. The control circuit uses a thermostat or pressure control to sense the condition in the refrigerator and then sends a signal to a remotely operated electrical switch that controls compressor operation. This remotely operated electrical switch can be either a contactor or a motor starter.

CONTACTORS

The contactor contains a coil and one or more switches. When current flows through the coil, as a result of a thermostat or pressure control sending an electrical signal, an electromagnetic force is created that causes the switch(s) to either open or close. A set of contacts, or switch, that is open when the coil has no current flowing through it, is called a normally open contact, abbreviated NO. If the contact is closed when no current is flowing through the coil, it is called a normally closed contact, abbreviated NC. Any number of switches, or sets of contacts,

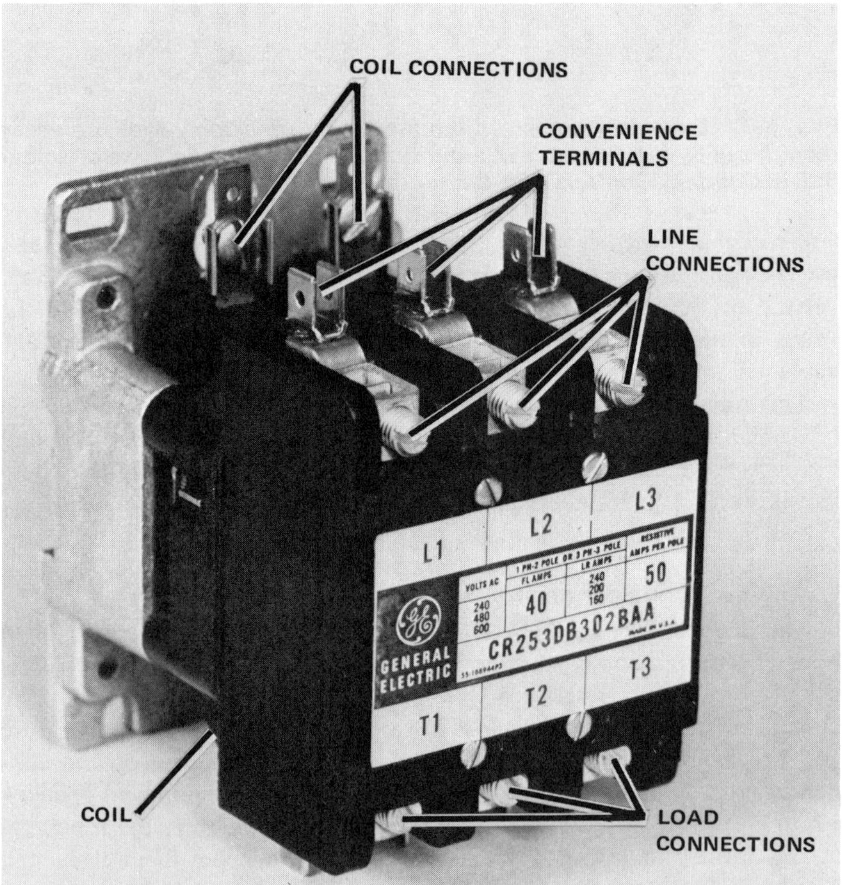

FIG. 8-8. The name plate of this three-pole contractor shows that it has an inductive rating of 40 amps, a resistive rating of 50 amps. The coil rating, not visible here, may be found on the coil itself. *Courtesy General Electric Co.*

can be incorporated into a contactor. Contactors with one, two, three and four sets of contacts are quite common; those with five and six sets are not as common. The term *pole* is often used in place of the term contact or switch. The term *4-pole, NO contactor* refers to a contactor with four switches that are in the open position when current is not flowing through the coil.

The service technician should be aware of how to specify a replacement contactor because it is a very commonly replaced component. In addition to specifying the number of poles and whether they are NC or NO, a capacity or current carrying capability must be specified. A contactor with an ability to carry 25 amps may well burn up, or at least have a shortened life, if consistently called upon to carry 40 amps. Typical increments of current rating are 15, 20, 25, 30, 40, 50, 60 and 75 amps. Going to the next higher rating will cause no problems, assuming the contactor will physically fit into the available space; going smaller, however, is definitely unacceptable.

Many contactors, Figure 8-8, will have two ratings, *inductive* and *resistive.* The inductive rating refers to loads, such as motors, that have an initial high current draw that drops back to full load current. A resistive load is an electric resistance heater, such as that used in defrosting applications. Usually, a contactor will have the ability to handle a higher resistive load than an inductive load. Matching the type and size of the load to the rating of the contactor will prevent grief at a later date.

A final consideration, in the selection and application of a contactor, is the coil voltage. Contactors are typically available with coils rated at 24v, 115v, 208v, 230v and 277v. In commercial refrigeration, the 115v, 208v and 230v coils are most common. A replacement contactor, identical in every respect to an original, will be useless unless the coil rating is also identical. A complete contactor specification will be: 3-pole, NO contactor with a 40 amp resistive, 30 amp inductive rating and a 230v coil. If you completely identify the contactor you want, in this manner, there is a good possibility that you will get what you need. *Always check to make sure you got what you asked for, and that you asked for what you wanted, before you leave a supply house.* That should rank somewere between *thou shalt not steal* and *thou shalt not kill* in your *rules to live by.*

MOTOR STARTERS

A motor starter, Figure 8-9, is nothing more than a contactor with built-in current overload protection. All previous comments about contactors apply to motor starters. The built-in overload protection deserves a few additional words.

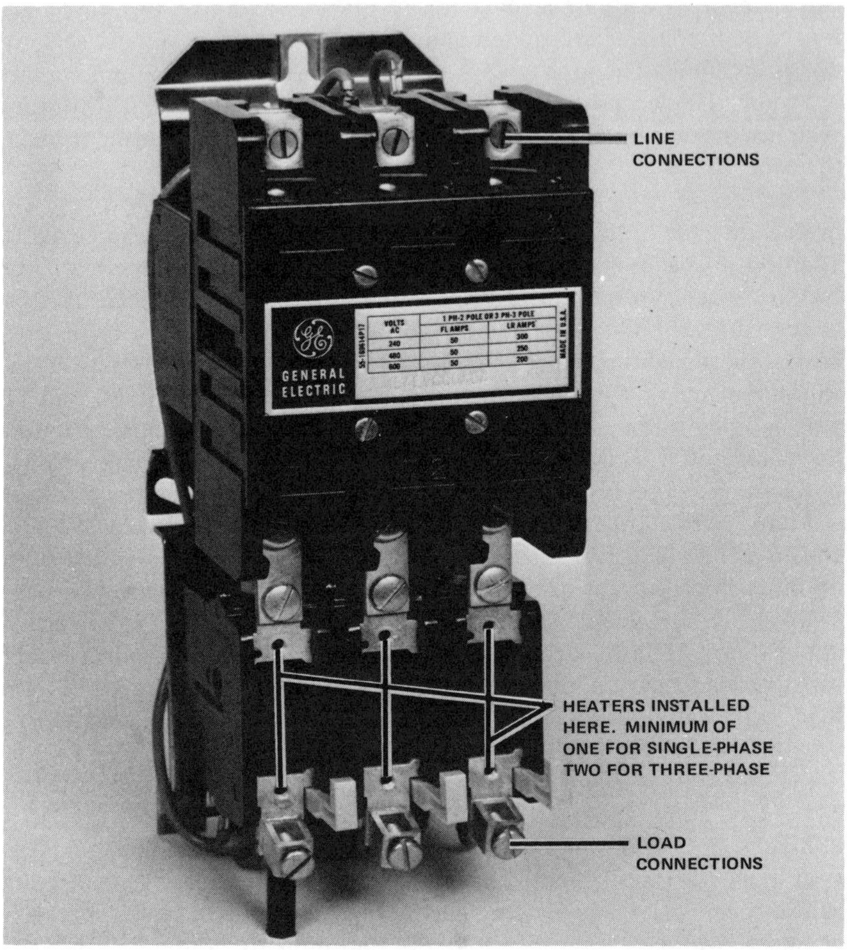

FIG. 8-9. This motor starter is essentially a contactor with a built-in motor protection device. *Courtesy General Electric Co.*

An electric motor is designed to do a particular job. In the course of doing the job, it will draw a current measured in amps. The maximum current it is designed to draw continually is called the *Full Load Amps,* abbreviated F.L.A. One characteristic of electric motors is that they are so accommodating that they will try to do any job they encounter. If they are asked to work harder than they are designed to do, they will make the effort and become overloaded, that is, will draw a current greater than intended. This usually results in the generation of excessive heat and a shortened, sometimes drastically shortened, life for the motor.

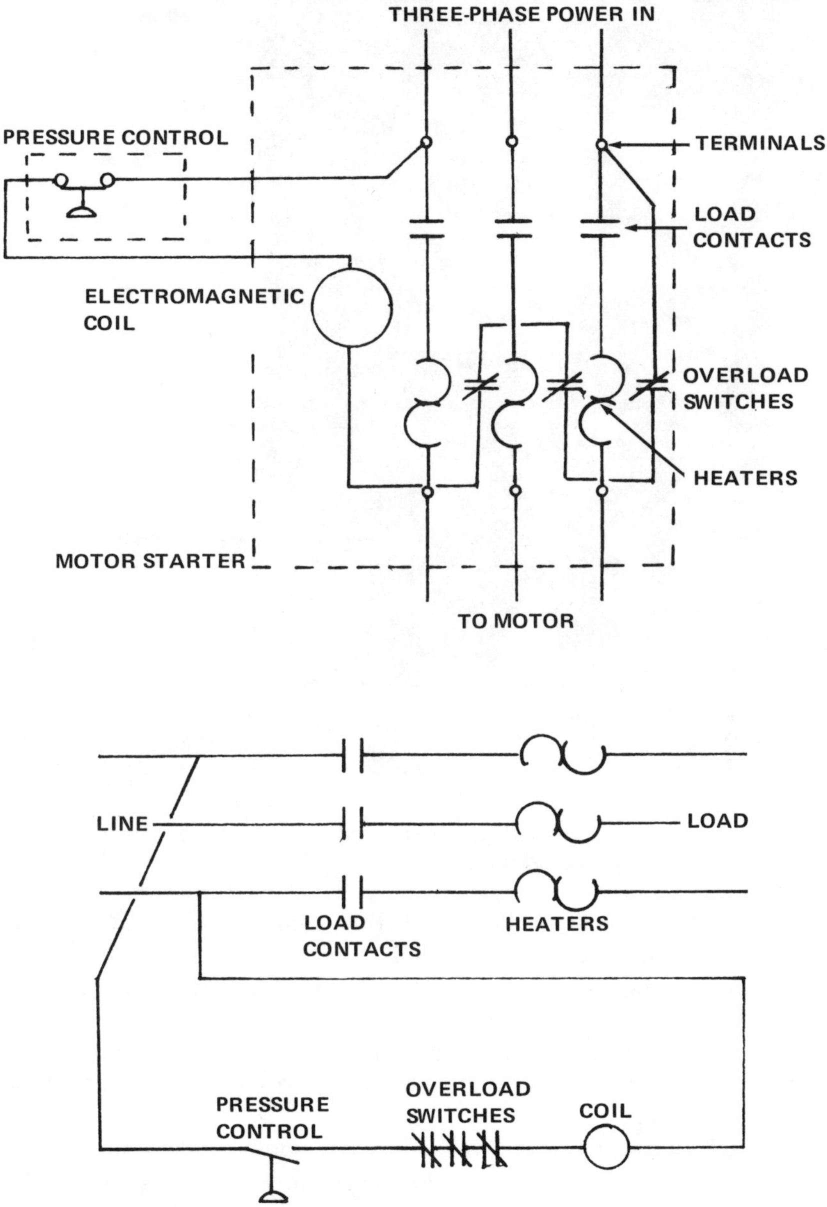

FIG. 8-10. Pictorial diagram, above, and schematic diagram, below, of motor starter with pressure control.

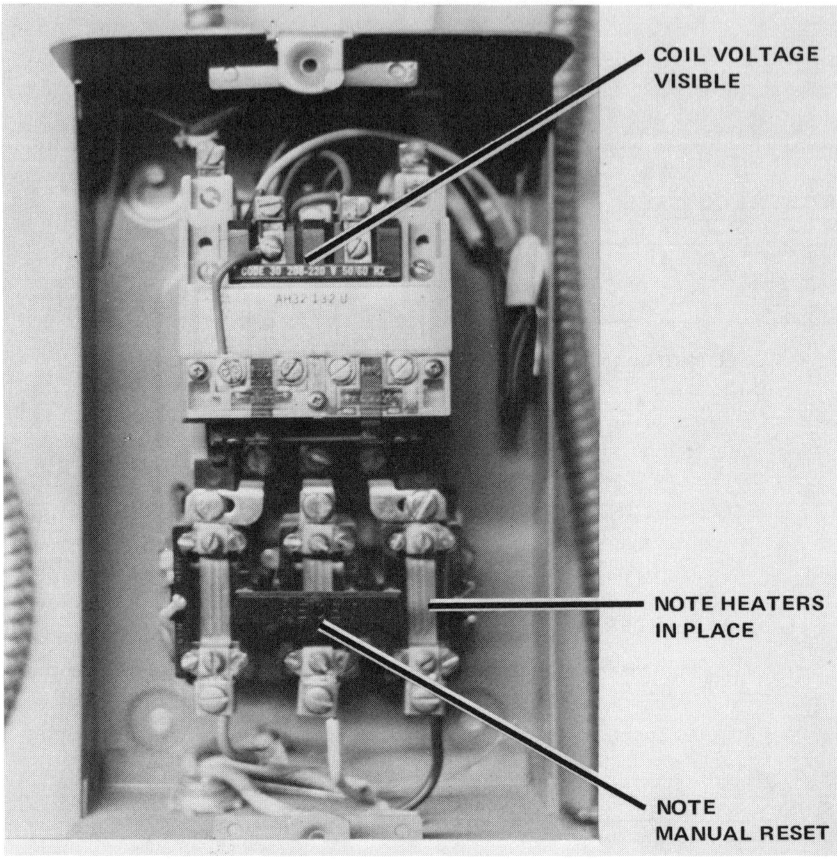

FIG. 8-11. Motor starter with heaters in place. Motor can be restarted after overload by depressing the reset bar.

Many motors, particularly those in hermetic and semi-hermetic compressors, have overload protection built into them. There are, however, many instances where the motor starter is used to back up the compressor protectors or to provide the first line of protection on open compressor applications. There are enough motor starters used in refrigeration applications to justify the serviceman knowing how they work and how they can be serviced.

The overload protection is provided by two components, a heater and a switch. The heater is in series with the motor. The current that the motor draws flows through the heater and heat is generated. If the current draw becomes excessive, the heat generated becomes excessive. Now the switch comes into play. The excess heat causes the switch, which has a bimetal sensing element, to open. The switch is *not* in series with the motor, but rather is in series with the *coil*, Figure 8-10. When the switch opens, the coil no longer has current flowing through it and the contacts will open. That, in a nutshell, is how the motor starter works.

The overload protection offered by a motor starter can be matched to the electrical load. When a starter is purchased, it has a nominal rating based upon contactor design, as previously described. The current that will cause the switch to open, however, is dictated by the heater selected. There may be as many as a dozen different heaters that can be used with one size motor starter. Each heater will cause the switch to open at a different current draw. An important part of specifying a motor starter is the specification of the heater to go with it. Usually, by indicating the F.L.A. of the load, you will get a heater which will cause the starter to shut down the motor when current is 10% higher than maximum.

Motor starters are available with manual Start-Stop buttons or they can be wired directly into a control circuit. Typically, the starter that has tripped out on overload must be manually reset, Figure 8-11. It tripped out for some reason and that reason must be determined by the serviceman before the system can be put back into permanent service. Unfortunately, that is very often easier to say than to do, as we'll see in a little while.

SERVICING WALK-IN REFRIGERATORS

Servicing the vapor compression cycle has been pretty well covered. The beauty of this refrigeration cycle is that, once it is understood for small systems, the principles and techniques can be extended to larger systems. It might be worthwhile, at this point, to explore the most dreaded happening of them all, the compressor burn-out.

COMPRESSOR BURN-OUTS

The term burn-out pretty much says it all. Full hermetic and semi-hermetic compressors have their motors within the same housing as the compressor mechanism. The electrical windings must be cooled and, in hermetics, the cooling is done by suction gas passing over the windings. If there is no refrigerant, or low refrigerant, there will not be enough vapor to cool the compressor. Similarly, if heat is generated by a bearing failure or if condensing pressures are unduly high, a breakdown in the insulation covering the windings can be expected. A breakdown results in a short circuit and still more heat is generated. The additional heat causes the oil and refrigerant to break down, creating sludge (carbon mixed with oil) and acid. You can begin to see a grim picture developing. Eventually, everything comes to a grinding halt, usually because some protective device, either an internal overload or a motor starter or circuit breaker, opens the circuit.

Since, in a typical walk-in refrigeration application, there are many dollars tied up in products stored in the box, it is imperative that the system be put back into operation quickly. This means changing the compressor. It may or may not be rebuilt later, but the main problem now is to change the compressor and, equally important, to do what is necessary to prevent another burn-out in the future.

There are three generally recognized degrees of severity of a burn-out. With a mild burn-out, the motor windings are open or shorted or grounded, but there is little or no sludge or acid. With a medium burn-out, the windings are open and some sludge and acid are formed, as indicated by discoloration of the compressor oil and a slight pungent odor of the refrigerant (whatever pungent means). With a severe burn-out, there is a massive formation of sludge and marked formation of acid, as indicated by heavy deposits in the discharge and suction lines and a sharp, acrid (see pungent) odor from the refrigerant. You'll know when you have a severe burn-out. The smell will knock you on your duff.

Mild Burn-Out

A mild burn-out only requires that the compressor be replaced, that a new liquid filter-drier be installed, and that a good vacuum be drawn on the system prior to charging it. If there is such a thing as a good burn-out, this is it.

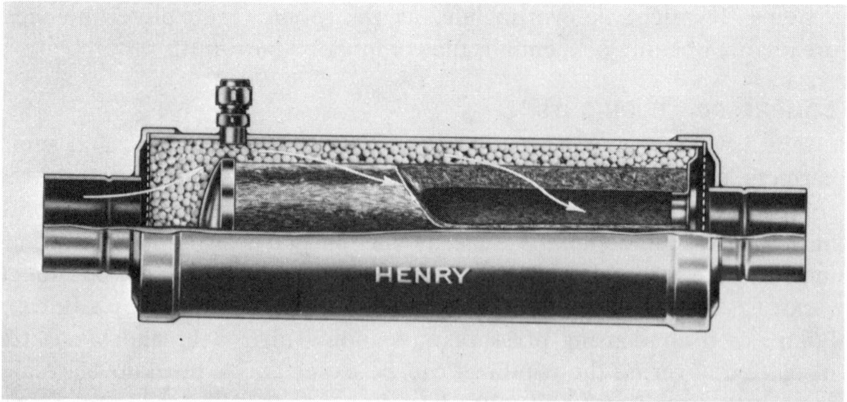

FIG. 8-12. Suction line filter-drier protects the compressor from acid contaminants after a burn-out. *Courtesy Henry Valve Co.*

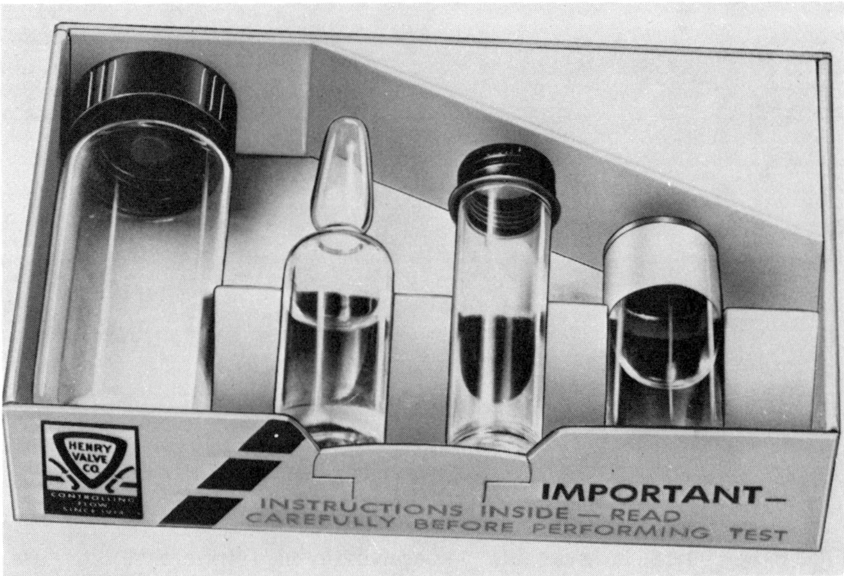

FIG. 8-13. Acid test kit. *Courtesy Henry Valve Co.*

Medium Burn-Out

The medium burn-out can be handled like the mild burn-out, with the addition of a suction line filter-drier, Figure 8-12. A suction line filter-drier is designed to trap any contamination in the system and prevent a recurrence of the burn-out. The installation of the suction line filter-drier will be all that is usually required. How can you be sure?

The *acid test kit,* Figure 8-13, is a valuable tool for assuring that a burn-out will not recur. Any acid left in a system will have a tendency to attack the windings of a new compressor. This attack will succeed, although the time needed will depend upon the strength of the remaining acid. With the acid test kit, you draw a sample of oil from the system and test it for acid content. The test costs a couple of dollars, and takes a couple of minutes, but it helps you to sleep at night without fear of the burn-out recurring.

Once the system has been reassembled and back into operation, it should be run for about 24 hours to allow the suction line and liquid line filters to do their stuff. If the *acid test* still shows traces of acid, a replacement set of driers should be installed. Another 24 hours, and another acid test, should tell the tale. If the burn-out was indeed a medium one, it shouldn't take more than one or two filter changes to get a clean system.

FIG. 8-14. R-11 is available in conventional refrigerant cylinders. Opening the valve, with the cylinder inverted, will force pressurized R-11 through the system to be cleaned.

Severe Burn-Out

In a severe burn-out, large amounts of sludge and acid are formed and are present throughout the system. This contamination should be cleaned out before a new compressor is installed. Refrigerant-11, in pressurized cylinders, Figure 8-14, is used as a solvent to flush out the system. R-11 liquid is forced through the discharge line, condenser and liquid line. Usually the TXV is removed from the system. Then the R-11 liquid will come out of the liquid line at the valve location. R-11 can then be forced through the evaporator and suction line, cleaning the low side of the system. The R-11 is generally allowed to flow until it comes out clean. It is a good solvent and this doesn't take too long.

After the new compressor is in place and the TXV, after having been thoroughly cleaned with R-11, is again installed, new liquid line and suction line driers should also be installed. Pull a good vacuum on the system, Figure 8-15, break the vacuum with refrigerant or nitrogen, and repeat twice more. Finally, charge the system and you are back in business. After 24 hours, use that acid test kit and make sure that all trace of acid has been removed and that the oil in the system is clean.

A good clean-up and suction filter protection are quite important. You can always come across stories about compressors being replaced annually until the system had a good clean-up. They're not tall stories,

FIG. 8-15. Vacuum pump used to evacuate refrigeration systems. *Courtesy Watsco, Inc.*

but rather the factual evidence of what will result when too few precautions are taken in replacing a burned-out compressor.

ELECTRICAL PROBLEMS

A typical walk-in refrigeration system has one or more evaporator fans that run constantly, a compressor and one or more condenser fans that are energized by a contactor or motor starter, a control circuit with a thermostat or pressurestat, possibly internal compressor protection, and perhaps an oil safety switch in series with the coil of the contactor or motor starter. Let's look at some of these components from the point of view of the troubleshooter.

COMPRESSOR

The techniques described in Chapter 6 for single-phase compressors in reach-in systems are also pertinent to single-phase compressors in walk-in systems. Three-phase compressors are even easier to troubleshoot because starting components, such as a relay and capacitor, are not used. The key to a properly functioning three-phase compressor is the integrity of the motor windings.

In checking for windings that are shorted or grounded, an ohmmeter is typically used. The standard ohmmeter has a battery capable of developing a potential of perhaps 1½ volts dc. It is quite possible for such a meter to indicate good windings when, in fact, a short is present.

A *megger* or megohmmeter is capable of developing substantially higher potentials than can be obtained with a standard ohmmeter. Various designs are available that can produce several hundreds of volts dc. A potential of that magnitude will quickly show weak compressor windings.

CONTACTORS

Problems with contactors fall into two general categories: first, there may be a problem with the contacts themselves; second, the coil may be defective.

If the contacts do not pull tightly together, arcing may occur that will eventually cause the contacts to weld together. When this happens, the compressor may run continuously, even though the coil is de-energized. A good test for any contactor is to measure the voltage on the line side and on the load side of the contact. The voltage readings should be identical. Any significant decrease in voltage across the contacts signifies that resistance is present in the form of an air gap or perhaps corrosion. This is bad for the equipment and bad for the contactor.

The coil itself may burn out, a not uncommon problem. A simple continuity test will usually tell the story. Burn marks and a distinctive burned odor are also indications that the coil is defective. A proper voltage reading across the coil, accompanied by a failure of the contacts to pull in, is a good indication of a defective coil.

MOTOR STARTER

Since a motor starter is a contactor with the addition of circuit protection, you might expect the information about contactors to be applicable here. The fact is that it is, the only difference being problems associated with the heater and overload switches.

A starter that continually cuts out on an overload may do so for a number of reasons. The load may, in fact, be excessive and a careful checkout with an ammeter is necessary to insure that there is actually a problem with the starter. The heater may be the wrong size or defective. If the load is not excessive, this is a good possibility. Most likely, but certainly not impossible, is a problem with the safety switch itself. Generally, the coil, contacts, safety switch, and heater are available as repair parts and it is quite common to replace one or more of them rather than installing a whole new starter which can be rather expensive and usually unnecessary.

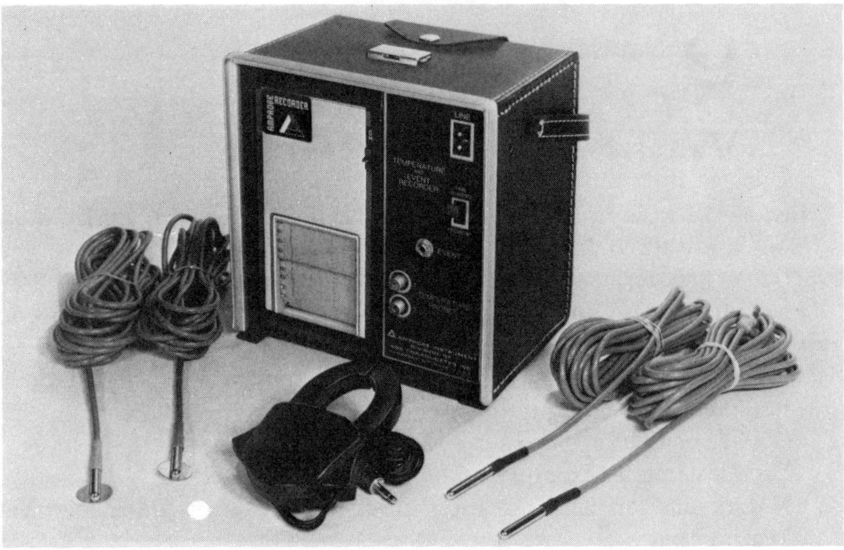

FIG. 8-16. This recorder can track temperature, voltage or current levels over extended periods of time and spot intermittent problems. *Courtesy Amprobe Instrument.*

Occasionally, intermittent electrical problems, either with the load or with the supply power, occur. These are not readily detected since they invariably occur when you are home sleeping. A recording voltmeter and ammeter, Figure 8-16, is helpful in monitoring such problems 24 hours a day. A momentary loss of power, in the wee hours of the morning, could very well result in a safety trip and you would never know it unless you had a recording instrument monitoring it. As the problems get more subtle, the methods needed to detect and solve them become more sophisticated, and the tools more expensive.

9

Walk-in Freezers

Just as the walk-in refrigerator compares to the reach-in refrigerator, so does the walk-in freezer compare to the reach-in freezer. Its prime function is long-term storage at temperatures in the neighborhood of 0°F.

The refrigeration and electrical systems are quite similar to walk-in refrigerators with the following exceptions:

1. While R-12 is commonly used, some equipment will use R-22 and an increasing number of units employ R-502.
2. Hot gas or electric defrost cycles must be included.
3. Condensate must be prevented from freezing in the box.
4. A door mullion heater is often installed to prevent the door from freezing shut.

The refrigeration and electrical systems are quite similar to reach-in freezers except:

1. Externally equalized TXV's are usually used.
2. Electrical control circuitry is more involved, due to the increased size of the equipment.
3. Oil pressure safety switches are often employed on larger compressors.
4. A liquid/suction heat exchanger or similar device is often used.
5. A crankcase pressure regulator may be used.

CONFIGURATIONS

There is little to distinguish a walk-in freezer from a walk-in refrigerator. Since the temperature of a walk-in freezer is considerably lower than the typical refrigerator, it would seem advantageous to use a thicker wall to reduce transmission heat gain. This was often the case in older boxes using fiberglass and cork insulation. However, current *foamed-in-place* construction, being so efficient, often makes the added cost of thicker panels impractical when compared to the slight gain in energy savings.

Far more important in freezer construction is structural integrity, which is kind of a cute way of saying that the box must be tight. Air leaks in a walk-in refrigerator usually result in a bit more *on-time* for the compressor. In a walk-in freezer, however, air leaks lead to ice formation. A box with leaky roof seams will, in short order, develop stalactites of ice due to the condensing and freezing of the humid air leaking into the box. Openings in side walls become filled with ice and

FIG. 9-1. Accumulated ice around an improperly-fitted door of a walk-in freezer.

lovely ice waterfalls develop: esthetically delightful, but practically a mess, Figure 9-1.

Mullion heaters are quite frequently found around the door frames. They keep the doors from freezing closed, which is quite possible with long-term storage. They also keep moisture from collecting around the door frame and sitting there in the form of droplets. This is usually more of a problem with metal boxes than with the wooden type.

CONDENSATE DISPOSAL

Disposing of condensate, removed from the air, is not a problem in a refrigerator. Since the temperature of the air in the box is above freezing, the water will run from the coil, into a pan, and from there into a drain leading from the box. The condensate can be directed to a sanitary drain, to an electrically heated disposal pan, or to a condensate disposal pump for eventual pumping to a remote drainage point.

Removing condensate from a box where the air temperature is below freezing is a bit more involved. During the defrost cycle, the coil surface is heated to above freezing, the frost on the coil melts, and the condensate drips into a pan below the coil. If the pan is at a freezing temperature, the condensate will immediately refreeze and remain in the box. For this reason, the collection pan beneath a low temperature coil must be heated,

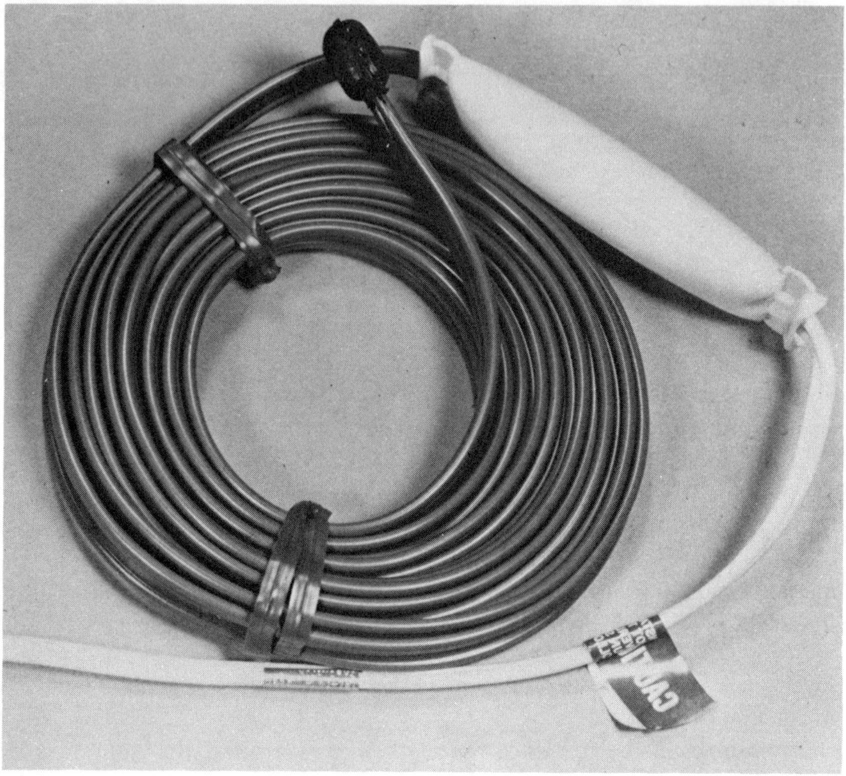

FIG. 9-2. Ribbon heaters keep the condensate drain line from freezing.

at least during the defrost period. In the case of electric defrost systems, the heater will be mounted in the pan or immediately below it. In a hot gas defrost system, the hot gas is routed through tubing bonded onto the bottom of the pan, or lying in the pan, before it passes into the evaporator. This added heat will keep the water flowing into the drain line. At the termination of the defrost cycle, the heater is de-energized or, in the case of hot gas defrost, the hot gas stops flowing.

Once the water is in the drain line, it may still freeze if the drain tube is below freezing. Since the drain is usually field installed, the installer must be aware of the need for a heater, of some sort, to keep the drain tube above freezing. Typically, ribbon heaters, Figure 9-2, are used for this purpose. Since ribbon heaters are available for several voltages, it is important to match the heater to the voltage of the power supply. If the drain tube has an adequate pitch and the proper ribbon heater is installed, there should be no problem with condensate flow.

Failure of the condensate removal system is not uncommon and the results are usually spectacular. The people using the equipment are probably unaware that something may be amiss until one day they notice a huge ice build-up in the drain pan that sometimes spreads up over the coil. Many times the ice build-up has become so severe that the fan blowing air across the coil is prevented from turning. A burned-out fan motor may result from something as insignificant as a frozen drain line.

REFRIGERATION SYSTEM

Most of the freezer equipment in use today uses R-12. It is not unusual, in older equipment, to see R-22 being used and, quite frequently, newer units will use R-502. Before any meaningful system analysis can be undertaken, it is necessary to identify the refrigerant in the system. Usually, a nameplate on the unit will indicate what refrigerant is being used. Find that identification plate and you're on your way. What's that? You say you can't find any indication at all? Fortunately, all is not lost.

Since there is a very specific relationship between refrigerant pressure and boiling temperature, we should have no difficulty identifying the refrigerant as long as the system isn't totally empty. With the system operating, measure the low side pressure and the temperature of the refrigerant on the evaporator side of the TXV. Refer to your refrigerant pressure/temperature chart and find the temperature you measured. Move across the chart until you see a pressure close to that you measured. Looking up to the head of the column should give you the refrigerant in the system.

Suppose you measured a pressure of 21 psig and a temperature of 20°F. Moving across the chart from the 20°F temperature reading, shows a pressure of 21 psig for R-12, about 43 psig for R-22 and about 52 psig for R-502. The refrigerant in our system would have to be R-12.

Suppose you measured a pressure of 29 psig and a temperature of about 0°F. Reading across from 0°F we see pressures of 9, 24 and 31 for R-12, R-22, and R-502. Our refrigerant would have to be R-502. Sometimes your readings may be perfect and sometimes there may be some variation. Generally, the characteristics of R-12, 22 and 502 are sufficiently different, so there shouldn't be any problem if you take your readings carefully.

If the system is totally empty, there are two things you can do. First, look at the TXV. It may have the refrigerant stamped on it or it may be color coded. Remember: Yellow=R-12, Green=R-22, Violet=R-502. Second, take down the model number, call the local distributor handling that make of equipment, and ask him. Finally, call the manufacturer when all else fails. If you have to do this more than once in your life, you deserve a spot in the Guinness Book of World Records.

SYSTEM PRESSURES

We've seen a way of predicting what the pressure will be in the high side of the refrigeration system. It is pretty much determined by the temperature of the air passing over the condenser and the refrigerant in the system. Low side pressures can be similarly estimated.

A good rule to remember is that the refrigerant temperature in an evaporator coil is usually 10 to 15°F less than the temperature of the air in the box. In a freezer, we expect a box temperature of about 0°F, so the refrigerant temperature would be about −10 to −15°F. Looking at our pressure-temperature chart, we see that an R-12 freezer will run about 2 to 4 psig on the low side, an R-22 unit about 12 to 16, and R-502 about 19 to 22 psig. Incidentally, we can apply the same rule to refrigerators. If the box temperature is 35°F, we would expect refrigerant temperatures in the evaporator of about 20 to 25°F. This means low side pressures in an R-12 system will be about 21 to 24, R-22 about 43 to 48, and R-502 about 52 to 58 psig. However, R-22 and R-502 are rarely seen in this application.

On start-up, since the box temperatures will be quite high, you would expect to see relatively higher pressures on the low side of the system. This is quite common, with both temperature and pressure dropping as the box pulls down to normal operating pressure.

CRANKCASE PRESSURE REGULATORS

The compressors used in low temperature applications are designed to handle low pressures on the suction side. Under typical low side pressures, they will run at a current draw approaching their full load rating. There are two conditions, however, when they may tend to run at an overload condition.

The first of these is under a start-up situation. The box temperatures are high, hence the rate at which refrigerant is evaporating in the evaporator is also high, resulting in elevated suction pressures. These higher-than-normal pressures result in high running amperage.

The second situation results after a defrost cycle is terminated or, in the case of a hot gas defrost system, during the defrost cycle itself. In both instances the evaporator is warm and, in the case of hot gas defrost, high pressure discharge gas is going directly into the low side of the system. In both cases, higher-than-normal operating pressures could very well shorten the life of the equipment, if precautions are not taken.

In these cases, a crankcase pressure regulator is installed in the suction line between the evaporator and the compressor, Figure 9-3. This device

FIG. 9-3. A pressure regulator in the suction line. The access port may be used to measure pressure in the evaporator. The service valve port may be used to measure suction pressure at the compressor.

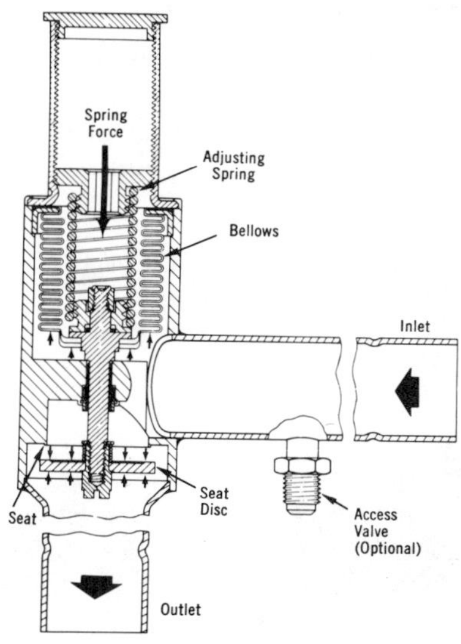

FIG. 9-4. The inlet and outlet of the suction pressure regulator are marked on the body of the regulator to assure proper flow. This spring-loaded device is preset at the factory, but may be field adjusted. *Courtesy Sporlan Valve Co.*

is designed to maintain a preset pressure on the compressor side of the control. It is a spring-loaded device, Figure 9-4, that is usually wide open when low side pressure is below the pressure setting of the control. As low side pressure increases, the regulator closes down, preventing suction pressure from increasing at the compressor. The pressure in the evaporator can become quite high, but the crankcase pressure regulator prevents this high pressure from getting to the compressor.

A common low side setting, in an R-12 system, is 15 psig. Upon start-up or during or after defrost, an evaporator pressure as high as 40 to 50 psig may occur. The regulator will hold the pressure back in the evaporator and the pressure at the compressor will not exceed 10 psig. The running amperage will therefore remain within design tolerance and overloading will not occur.

LIQUID-SUCTION HEAT EXCHANGER

The liquid-suction heat exchanger, Figure 9-5, is designed to increase both subcooling in the liquid line and superheat in the suction line. If you think back to the Mollier diagram presented earlier, you will remember that increased subcooling increased operating efficiency of the system because it decreased *flash gas*. The decreased flash gas resulted in a greater *net refrigeration effect*. The subcooling that takes place in the liquid line is a direct result of the cool suction vapor, heading back to the compressor, absorbing heat from the warmer liquid in the liquid line, Figure 9-6.

The suction gas, since it is absorbing heat, will have a higher superheat. This can be desireable from two points of view. First, under a low

FIG. 9-5. Liquid-suction heat exchanger. *Courtesy Bohn Heat Transfer Div., Gulf + Western Mfg. Co.*

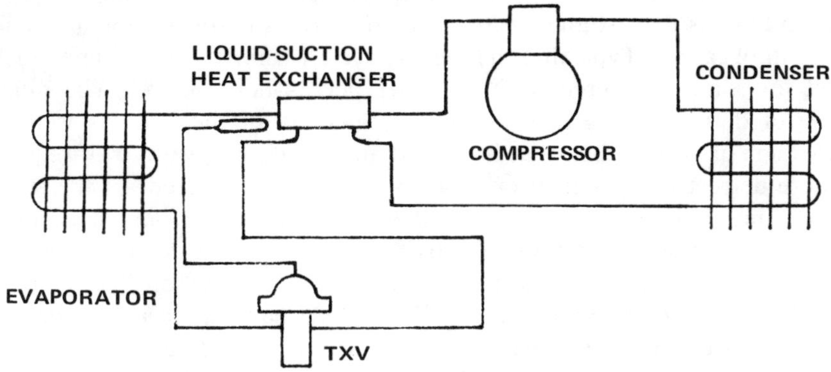

FIG. 9-6. Liquid-suction heat exchanger is installed between the TXV sensing bulb and the compressor.

load condition, there might be a possibility of liquid refrigerant getting into the suction line and being carried back to the compressor. The liquid-suction heat exchanger will cause some of this liquid to boil off into a vapor before it gets back to the compressor. Second, in a low temperature application, it is common for the refrigerant, even though it is superheated before it leaves the evaporator, to be at a temperature substantially below freezing. As an example: if the box temperature is 0°F, we might expect the refrigerant boiling temperature to be about 15°F below zero. If the expansion valve is holding 10° superheat, the refrigerant in the line would still be 5°F below zero. This will produce frosting of the suction line back to the compressor and even a build-up of frost on the compressor body itself. The liquid suction heat exchanger will heat up the suction vapor enough to bring its temperature above the freezing point, thereby eliminating the threat of a frosted suction line and compressor.

While it does seem that the liquid-suction heat exchanger is *the answer to a maiden's prayer,* actually there should be a element of caution used is employing one. Indiscriminate use could result in a compressor failure. Remember that hermetic and semi-hermetic compressors rely on suction gas cooling. If the suction gas is not sufficiently cooled, the compressor will run hot and either trip out on its overload, or worse, not trip out on its overload and have a drastically reduced life. It's a good idea to discuss the application of the liquid suction heat exchanger with a manufacturer's representative before applying it. After all, he is only a phone call away.

DEFROSTING

As was the case with reach-in freezers, walk-ins need a defrost cycle. Both hot gas defrost and electric defrost enjoy widespread popularity in this application. Typically, the defrost cycle is initiated by a time clock set for some number of defrost cycles per 24-hour period. The number of defrost cycles depends upon the usage of the box, as well as the time of year. A box that is opened and closed frequently will require more defrost periods than one that is infrequently entered. In the summer, the air is usually more humid than in the winter, and it is quite common for one or two extra defrost cycles to be set on the clock for this time of year. There is no pat answer to the question "How many cycles and of what duration should be set?" Usually, after a short period of time, it becomes apparent what the schedule should be.

HOT GAS DEFROST

A look at a wiring diagram for a typical hot gas defrost system, Figure 9-7, will give a good idea of how such a system is supposed to work. The timer is shown in the normal freezing position, with the circuit made

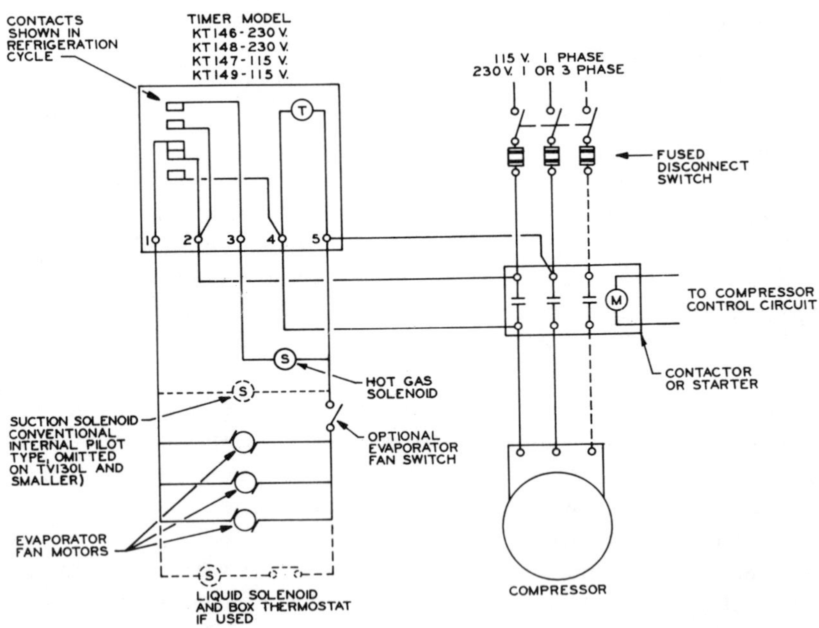

FIG. 9-7. Typical wiring diagram for a hot gas bypass defrost system. *Courtesy Kramer Trenton Co.*

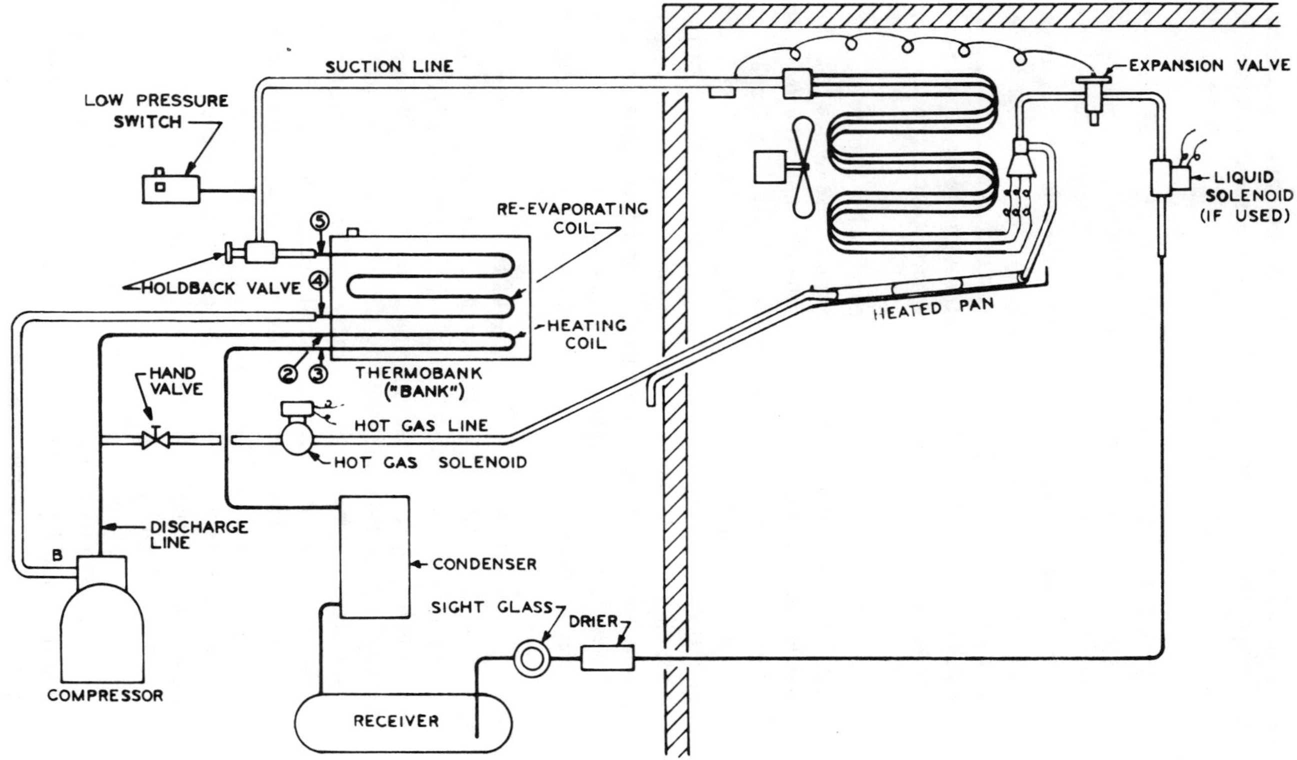

FIG. 9-8. Refrigeration system with a solenoid equipped, hot gas bypass defrost system and a Thermobank unit, *Courtesy Kramer Trenton Co.*

between contacts 1 and 2. Therefore, the evaporator fans are running. Note that the circuit between terminals 2 and 3 is open and that the hot gas solenoid is not energized. Operation of the compressor is independent of the time clock, being controlled by a thermostat designed to maintain box temperature.

When the time clock initiates the defrost cycle, contacts 1 and 2 open and contacts 2 and 3 close, de-energizing the evaporator fans and energizing the hot gas solenoid valve. Hot gas, flowing through the evaporator coil, will heat up the freezer quite rapidly if the fans are not shut down.

Figure 9-8 shows the refrigeration circuitry for this hot gas defrost system. The hot gas takes the path of least resistance, through the solenoid valve, bypassing the condenser. It flows through the drip pan heating circuit and then into the evaporator itself where it melts the ice from the coil.

One characteristic of hot gas defrost, compared with electric defrost, is that the cycle is usually substantially shorter in the case of hot gas. The flood of hot gas entering the evaporator very quickly melts the frost. In fact, it had better melt the frost very quickly because, the longer the cycle lasts, the cooler the discharge vapor becomes. Since cooling is not being done in the evaporator, the compressor does not have to work very hard. Since the hot gas bypasses the condenser and thermostatic expansion valve, there is little resistance to flow and the high side pressure tends to drop off. Typically, a hot gas cycle lasts for about 10 to 20 minutes, while electric heating cycles can last as long as 40 to 50 minutes or longer in extreme cases.

The defrost cycle will terminate when one of two things happens. In Figure 9-7, defrost is terminated by the timer. After the defrost cycle has run for the number of minutes set on the timer, the contacts switch back to normal cooling operation. An alternate method, used with some systems, is to have a thermostat mounted on the evaporator coil to sense that the coil surface temperature is increasing. When the temperature hits the set point of the thermostat, the defrost cycle is terminated. The timer is particularly valuable in hot gas systems because, if the hot gas cools down too much, as a result of low discharge pressure, the unit could remain in defrost indefinitely. The coil might never warm up enough to cause the defrost terminate thermostat to trip.

After the clock terminates the defrost, the hot gas solenoid will close and a normal freezing cycle will begin. The evaporator fans will remain off for several minutes. The coil is quite warm, as a result of hot gas flowing through it. If the fans came on immediately, the heat from the coil would be blown into the box, possibly causing some food spoilage or at least increased box temperature. The fans will remain off until the coil becomes cold. When it is cold enough, another thermostat, mounted on the coil surface or occasionally on a coil return bend, will close and start the fans. The cycle is now complete.

PROBLEMS WITH HOT GAS DEFROST

Hot gas flowing into a cold evaporator has a tendency to condense into a liquid. The liquid will then flow through the suction line back to the compressor, creating a situation that could very well lead to compressor failure. To counteract this possibility, the liquid refrigerant in the suction line must be re-evaporated. Two techniques are commonly used to aid this re-evaporation.

The *Thermobank,* Figure 9-8, manufactured by Kramer-Trenton, is a popular system for preventing flood back. During the normal freezing cycle, a reservoir of water is heated by hot discharge gas passing through a coil immersed in the reservoir. The suction line also passes through this reservoir of warm water. During the defrost cycle, any liquid formed in the evaporator is boiled off in the *Thermobank* and passes into the compressor as vapor.

A second device is the suction accumulator which is nothing more than a tank placed in the suction line. Liquid refrigerant, from the evaporator, falls into the bottom of the tank. The suction line draws vapor off the top of the tank, permitting the liquid to remain behind until it has vaporized. Occasionally, a heat source is provided to speed the rate of re-evaporation in the accumulator.

The technique used is not important, as long as it works. It is important that liquid not get back to the compressor. In designing a system for installation in the field, it is always good policy to follow the manufacturer's instructions or, in the absence of any instruction, to contact him directly for guidance.

A second possible source of woe, with hot gas defrost, is low ambient situations. If an air cooled condenser is too efficient, because the outside air is too cold, then low discharge pressures will result. This is not bad, up to a point. Generally, a pressure drop of about 100 psi across the expansion valve is desireable in an R-12 system. As the pressure drop decreases below this point, the efficiency of the system suffers. This is particularly true when the system goes into defrost. The hot gas may not contain enough heat to do an adequate job of defrosting. The result is unusually long defrost cycles or possibly incomplete defrost.

Increasingly, equipment is being produced with some sort of high side pressure control device, such as dampers operated by high side pressure that have been used on packaged equipment. Low head pressure modulating valves have also been developed and applied in such situations. Any field expedient can be used, as long as its effect is to increase the ambient temperature of the air passing over the condenser. A not uncommon solution is to build a small enclosure around the condensing unit that is almost completely sealed off in winter and wide open in summer. This is not too scientific, nor is it too precise in its result, but in many instances it will provide sufficient control and solve the problem.

ELECTRIC DEFROST

Electric heaters mounted on the evaporator coil have become quite popular as a means of removing accumulated frost. The wiring schematic for such a system is similar to the hot gas system. Again, a time clock initiates the cycle, Figure 9-9. In this instance, the fan motors on the evaporator are again de-energized and, in addition, the compressor is shut down. Simultaneously, the electric heaters are energized. One characteristic of this system is that it is of longer duration than the hot gas method. After a number of minutes, the frost will be seen to be melting and condensate will run down into the drain pan.

The cycle can be terminated in any of three ways. First, there may be a termination thermostat similar to that used in the hot gas method. An obvious precaution is not to have this thermostat influenced by the heater, but rather have it sense coil surface temperature. When the surface temperature increases, indicating the ice has all been removed, the cycle will terminate. Second, the timer itself may terminate the cycle. While it will invariably be used as a *fail-safe,* in the event that a termination thermostat malfunctions, it may very well be the only means of

FIG. 9-9. Time clock used to control the electric defrost cycle.

problems caused by their condensate.

From the other side: Manufacturers of the higher efficiency gas furnaces generally did not believe that conventional average-efficiency furnaces were soon to be closed out of the market — not by mandated efficiency numbers arrived at just because of the higher levels now being reached.

But while the state-of-the-art in heating could be viewed at the show, there was still another segment much in evidence which balanced the scales — that well-received segment which offered wood-burning, coal-burning, wood/coal burning (ad infinitum) stoves, furnaces, and boilers.

The market, it would seem, is not quite ready yet to abandon convention and support strictly

Stack heat rec conserve fuel,

BLOOMINGTON, Minn. — Heat reclaiming devices for chimney stacks can be practical energy savers, according to Paul B. Stegmeir, a consultant and lecturer on energy alternatives and conservation.

Stegmeir said his interest in the devices was rekindled by the findings of the University of California's Lawrence Berkeley Laboratories and a DOE-sponsored survey by Battelle Laboratories and the American Gas Association for Brookhaven National Laboratories.

Those research findings emphasized the problem of chimney stack heat loss above household furnaces and pointed to stack heat reclaiming as an

Solid state system for refrigeration

Defrost cycles staggered to a

AT THE SHOW — Defrost efficiency remains a nemesis in commercial refrigeration systems and is a target area for improvement.

Among defrost refinements at the show was Paragon Electric Co.'s microprocessor-based EC800 Programmable Defrost Controller.

The solid-state EC800 has programming capability for scheduling up to 96 defrosts per day, and can automatically stagger defrost starts (by 15 minute intervals) to reduce electrical peaks and "avoid overloading equipment defrost capacity," said Larry Wojcik, the company's product manager.

The unit has battery-powered carryover, enabling it to maintain the program in case of power outage, said Wojcik. Further, each output circuit has provisions for defrost termination as controlled by remote thermostat or pressure switch.

The unit also "has been designed to accommodate proportional defrost" based on humidity, said the company. This feature, to be available later this year, will allow intervals between defrosts to lengthen when humidity is low, and shorten when it's high, explained Wojcik.

ASHRAE

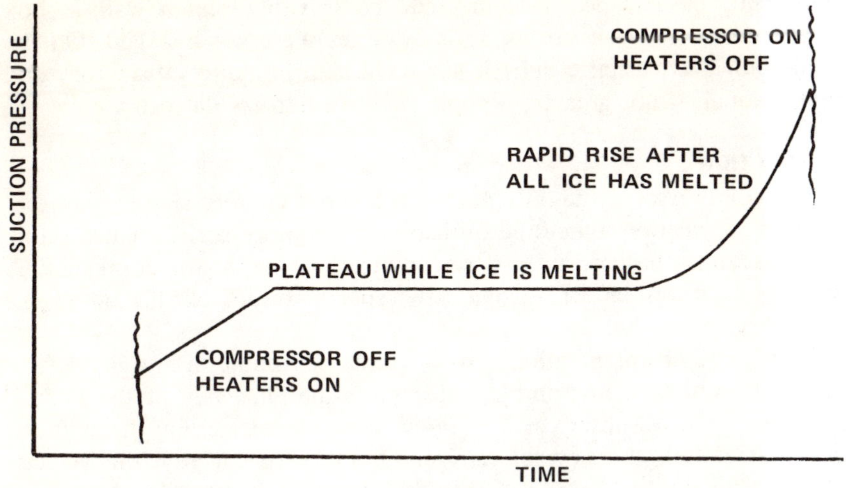

FIG. 9-10. Variations in suction pressure during the defrost cycle.

termination in the cycle. Finally, a pressure switch may be sensing suction pressure. This is kind of an interesting termination technique.

The pressure in the low side of the system, during the off cycle or defrost cycle, is directly related to the temperature of the coil. As the coil is warmed by the heater, we might very well expect to see the low side pressure increase. Actually, there may be no increase in pressure during the early part of the defrost cycle, because the heat going into the coil is being used to melt the ice on its surface rather than to increase its temperature. When all the ice is gone, however, we would see a rather rapid increase in low side pressure, Figure 9-10. A pressure switch, contained in the time clock, will sense this pressure increase and terminate the defrost cycle.

Upon termination of the cycle, the compressor will come on but the evaporator fan will not, as was the case with the hot gas system. Of course, the electric heaters are de-energized. When the compressor has run long enough to reduce the evaporator coil temperature, the fans will then start up.

There are not many problems associated with electric defrost and, for that reason, its popularity is growing. It does consume more expensive energy than hot gas defrost, because the heat from the hot gas is otherwise exhausted through the condenser. The initial cost of the electric defrost coil is more than for a coil using the hot gas technique but, by the time you add the extra accessories needed to assure maximum efficiency with the hot gas cycle, the costs are quite comparable. A malfunctioning electric heat system could conceivably heat a walk-in box to dangerous levels, if proper safety devices were not installed. On the whole, however, electric defrost has proven to be quite satisfactory and enjoys equal, if not greater, acceptance than hot gas defrost.

LUBRICATION

The compressor is the only part of the vapor compression system that requires lubrication. Once the oil leaves the compressor, as it invariably does, it cannot be used until it returns to the compressor crankcase. A well-designed system will provide for speedy return of this oil to the crankcase.

Low ambient was mentioned as a source of trouble in hot gas defrost cycles. In addition, low ambient decreases the efficiency of the normal freezing cycle, resulting in decreased flow of refrigerant through the evaporator. It is the flow of refrigerant, more specifically, the velocity of the refrigerant in the tubing of the system, that causes oil to be carried back to the compressor. If flow rate, therefore velocity, is unusually low, then the oil may very well trap out in the evaporator or in some other part of the system. The result of this could very well be reduced lubrication of the compressor and shortened life, sometimes drastically so.

Two solutions to this problem are: high side pressure control and proper sizing of refrigerant lines. High side pressure control was previously described. Proper sizing of the refrigerant lines, particularly the suction line, is quite important. It is interesting to note that a suction line that is undersized will produce unduly high pressure drops and a significant reduction in capacity. On the other hand, suction lines should not be oversized because, while pressure drops may be decreased, refrigerant velocity may be reduced to the point of inhibiting oil return. Again, and its been said often before, consult the manufacturer's recommendations for sizing suction lines. If the line is to be ¾ inches OD, up to 25 feet in length, and ⅞ inches OD, if the length is greater than this, then do it. There is a reason for the specification: that is, oil return coupled with high efficiency.

Refrigerant oils easily mix with the commonly used refrigerants. This property is called *miscibility*. Under low temperatures, however, this miscibility decreases. It may be advisable to use an oil separator, Figure 9-11. This device, while not 100% effective, will reduce the amount of oil that is circulating through the system and therefore reduce the possibility of the oil accumulating in the evaporator or elsewhere.

The oil separator is installed in the discharge line, as close to the compressor as possible. As the hot gas, laden with oil, enters the separator, the oil falls to the bottom and eventually is returned to the compressor

FIG. 9-11. The inlet and outlet of this oil separator are marked, as is usual. The return oil port is connected to the compressor. *Courtesy A C & R Components, Inc.*

crankcase through a float valve and oil return line. The purpose of mentioning this device is not to induce you to install it on every low temperature system you encounter, but to point out its importance. If a system has an oil separator, it must not be arbitrarily removed from the system if it is malfunctioning.

OIL PRESSURE SWITCH

There are two types of lubrication used in compressors: splash type, in which oil in the compressor crankcase is splashed onto the moving parts by the crankshaft, and pressure or forced lubrication. Forced lubrication uses an oil pump that is either built into the compressor or mounted on the outside. Pressure of about 50 psig forces the oil through machined passageways to the points requiring lubrication. Oil pump pressure can usually be measured by attaching a gage to a port provided for this purpose. Since the oil is in the compressor crankcase, it is subjected to both the suction pressure on the system and the action of the pump. The pressure at the gage is actually the sum of the suction pressure and the oil pump pressure. As an example: if the system is running a suction pressure of 15 psig and the pressure read at the oil gage is 68 psig, then the actual output of the pump is 68 minus 15 or 53 psig. The

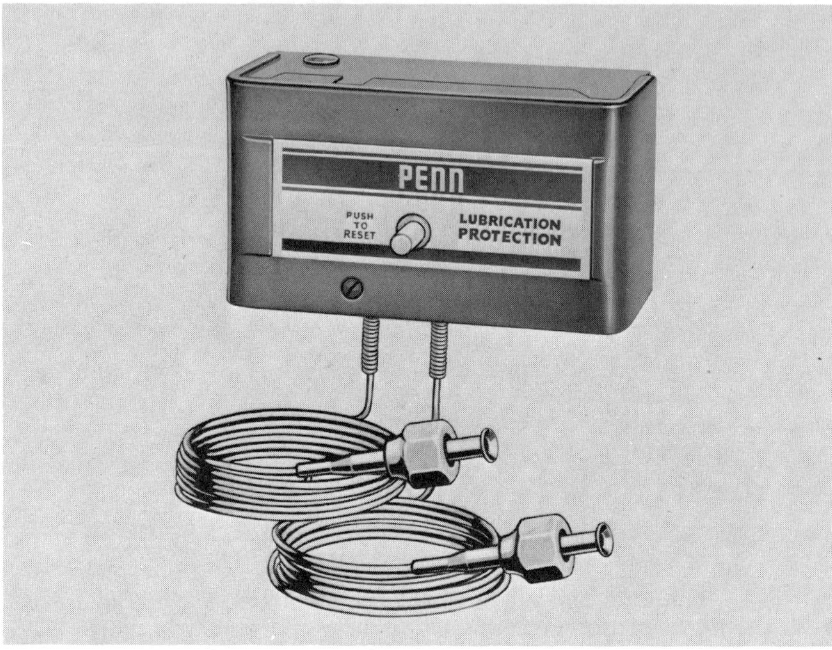

FIG. 9-12. An oil pressure safety switch. *Courtesy Penn Div., Johnson Controls, Inc.*

manufacturer's specifications will indicate what the oil pump pressure should be for the particular compressor.

An oil pump is provided on compressors typically used in larger walk-in refrigerators and freezers, in the range of several horsepower and up. To prevent a total failure of the compressor, in the event that an oil pump malfunctions, an oil pressure safety switch is used, Figure 9-12. This may be provided with the compressor or the manufacturer may recommend it as an option. If the decision is yours, to use or not use this safety device, use it.

The switch has a time delay, of about 60 to 90 seconds, built into it. If the oil pressure does not build up to a safe level within that time period, the compressor will shut down. If the pressure does build up in 60 to 90 seconds, the pressure switch will remain closed. A manual reset is usually provided, to prevent the compressor from cycling with inadequate lubrication and to signal you that something is wrong with the oil pumping system.

Just a word of caution. Under certain conditions, such as start-up after a fairly long off-period, the oil switch might cut out when, in fact, the oil pump is quite all right. It might happen that the oil has become logged with refrigerant. Upon start-up, the liquid refrigerant will boil out of the oil, causing the oil to foam. Since foam, passing through the oil pump, does not develop the same high pressure that oil alone will develop, the switch will trip out. By the time you reset the switch, all the refrigerant has boiled out, everything works normally, and head scratching time begins. If this happens at widely spaced, random intervals, it is not something to become concerned over. However, if it is a regular occurrence, further investigation is warranted.

CRANKCASE HEATER

If you suspect that the oil is becoming saturated with refrigerant at regular intervals, it may be time to check the crankcase heater or to consult the manufacturer and determine if a crankcase heater should be installed.

The crankcase heater is an electric heater that is strapped around the compressor crankcase, Figure 9-13. Other compressors are equipped with wells for the insertion of rod-type heaters. The heater is energized during the compressor off-cycle and warms the crankcase to prevent refrigerant vapor from migrating to the crankcase and condensing out into the oil. Refrigerant does nothing good when it is in the crankcase.

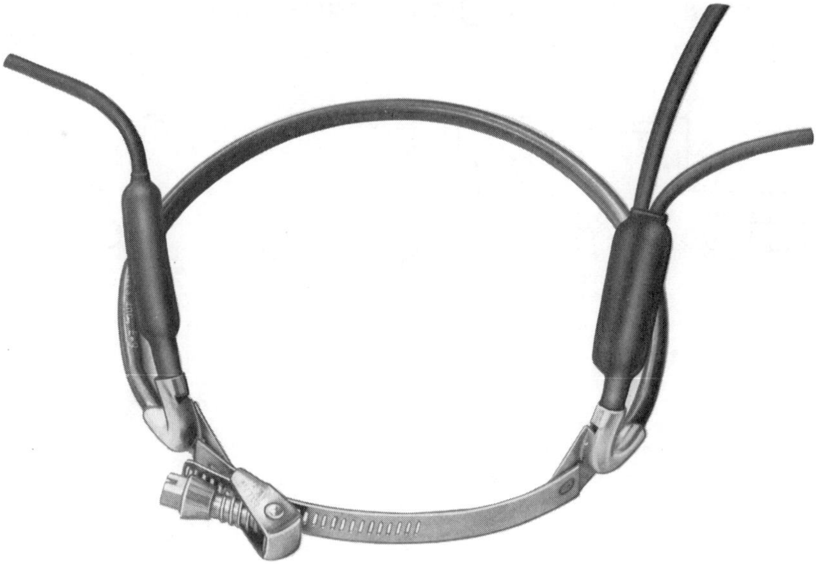

FIG. 9-13. Wraparound crankcase heater for a hermetic compressor. *Courtesy Heaters, Inc.*

On start-up, it causes oil foaming and may force a good amount of the oil out of the crankcase, resulting in insufficient lubrication. It also dilutes the oil to the extent that, even if oil is present, its lubricating characteristics are adversely affected.

The crankcase heater is a handy little device and is usually found on compressors exceeding several horsepower. It is particularly useful if the compressor is in a cold location because, in this situation, refrigerant migration is most severe.

THE PUMP-DOWN CYCLE

While the crankcase heater can be effective in preventing refrigerant from migrating to the compressor crankcase during the off-cycle, it is not the only remedy. The pump-down cycle is also very effective since it removes all liquid refrigerant from the evaporator, prior to the compressor shutting off. This prevents accumulation of liquid refrigerant in the evaporator, which is the coldest part of the system in a low temperature application. Such an accumulation could result in liquid slugging on start-up, that is, liquid refrigerant getting back to the compressor and causing all sorts of havoc.

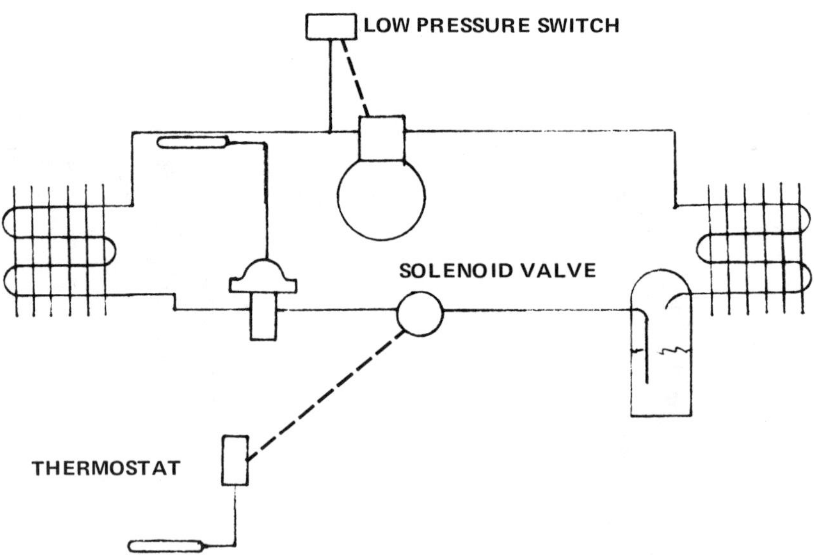

FIG. 9-14. System equipped to pump-down at the completion of a cycle. The solenoid is controlled by a thermostat and the compressor is shut down by the low pressure control.

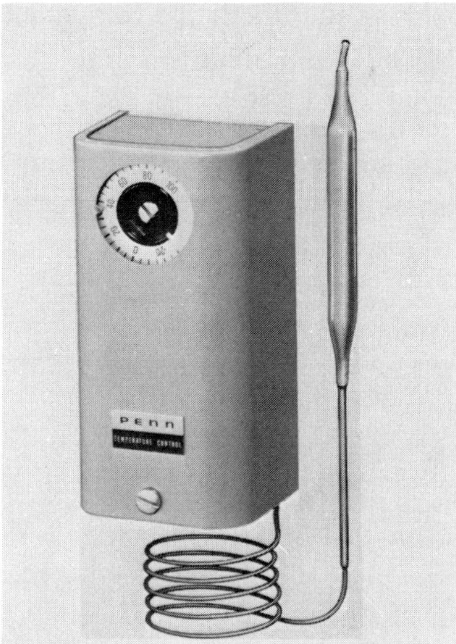

FIG. 9-15. Typical thermostat used to initiate pump-down at the end of a cycle. *Courtesy Penn Div., Johnson Controls, Inc.*

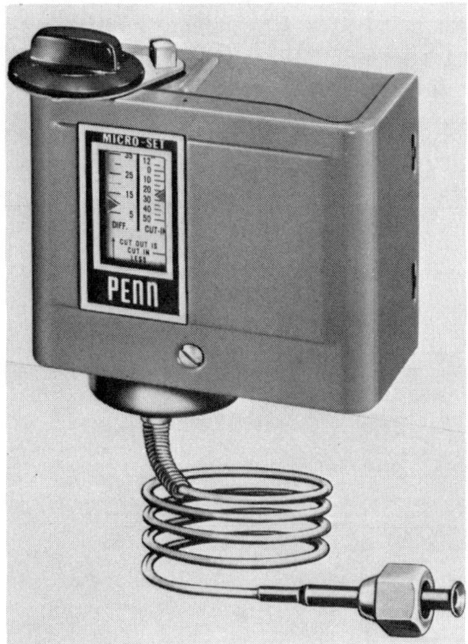

FIG. 9-16. Low pressure control is used to terminate the cycle after pump-down. *Courtesy Penn. Div., Johnson Controls, Inc.*

The pump-down system, Figure 9-14, employs a solenoid valve, that is controlled by a thermostat that senses box temperature, Figure 9-15. When the thermostat senses that the box temperature is at the set point, it de-energizes the solenoid and stops the flow of refrigerant. The compressor will continue to run and will pump the refrigerant out of the low side of the system. When all the liquid has been removed, the low side pressure will drop rapidly. A low pressure cut-out switch, Figure 9-16, is used to sense this pressure drop and to shut down the compressor.

When there is a call for cooling in the box, the thermostat will open the solenoid valve and permit liquid refrigerant to flow into the evaporator. As a result, low side pressure will rapidly increase, the low pressure switch will close, and the compressor will start.

In a system not using a pump-down cycle, the thermostatic expansion valve will be in a closed position after the compressor shuts off. This is the result of the low side pressure increasing, forcing the valve closed. It is this characteristic of the TXV that allows a pressure control to be used to control temperature in a refrigerated space. The use of the pump-down cycle, however, creates a very low evaporator pressure at shut down. This low pressure causes the TXV to remain open so that, when the solenoid opens upon demand of the thermostat, the liquid refrigerant will flow through the open valve directly into the evaporator.

The pump-down cycle is not restricted to low temperature applications, but is quite commonly used across the entire refrigeration and air conditioning spectrum. Its main reason for being is to prevent liquid slugging on start-up and to avoid dilution of the oil in the crankcase.

10

Water-cooled Condensers

All of the applications discussed to this point used air as the medium for condensing the refrigerant in the high side of the system. Actually, this use of air is a relatively new idea; early refrigeration machines being invariably water-cooled, Figure 10-1. The reason is not hard to determine.

In a Mollier diagram presented earlier, we plotted an air-cooled cycle. For a given low side condition, we could see that, as the head pressure increased, the heat of compression increased, and the *net refrigeration effect* decreased. With high head pressure, system efficiency decreases and it costs more to get the same cooling effect. In an air-cooled system,

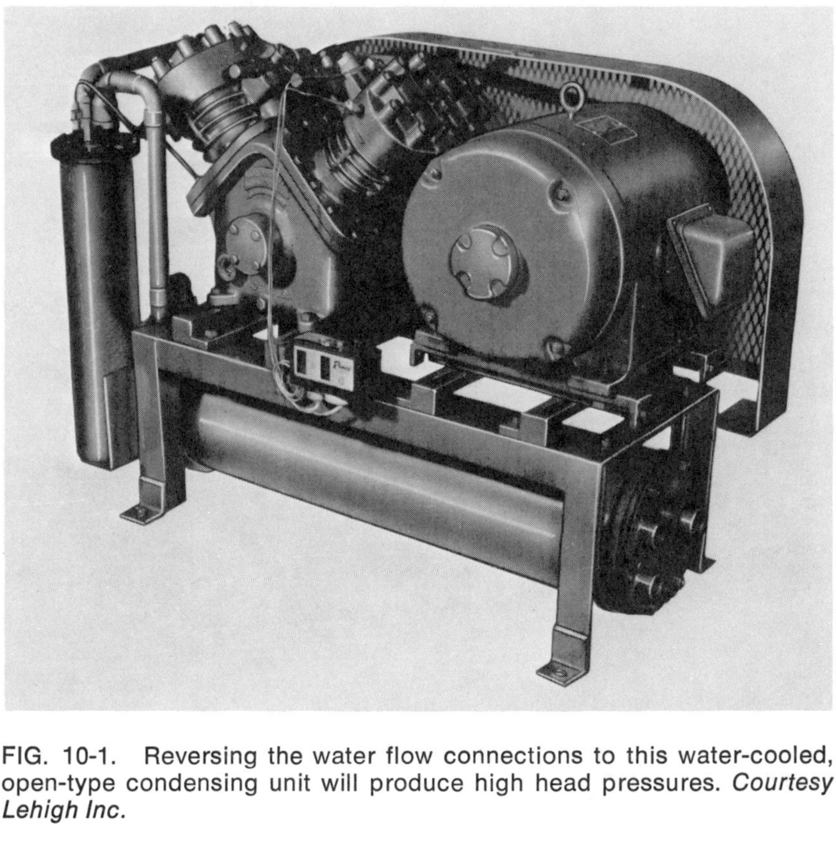

FIG. 10-1. Reversing the water flow connections to this water-cooled, open-type condensing unit will produce high head pressures. *Courtesy Lehigh Inc.*

head pressure is directly related to the temperature of the air passing over the condenser. The temperature of the air is dependent upon outdoor temperature, therefore, we can expect a drop in system efficiency as outdoor temperature increases. In the summer, when we most need refrigeration, we usually see a decrease in operating efficiency.

Water-cooled equipment is inherently more efficient than air-cooled. Water has better cooling performance than does air at the same temperature, therefore, less of it has to be moved over a heat exchanger and the heat exchanger can be a smaller size. An operating efficiency can be realized from this fact alone, Figure 10-2. The major benefit, however, is that high side pressure in the refrigeration system can generally be kept constant throughout the year, if the flow of water to the condenser is properly controlled. It is uniform head pressure, that results in maximum operating efficiency year-round, that makes water-cooled equipment highly regarded in the field.

Well, you ask, if it's that good, why isn't it used more often? And a fair question that is.

FIG. 10-2. The outlet from this water-cooled condensing unit is wrapped around the compressor case for added compressor cooling. *Courtesy Copeland Corp.*

LIMITATIONS

Perhaps the most obvious drawback to the application of water-cooled equipment has been the growing realization that water is not a limitless resource. Periodic water shortages, coupled with increased interest in conservation, has led to criticism of the water-cooled system. In the *old days,* water cooling was rarely questioned and was widely used. Today, more and more municipalities have laws limiting the use of city water for cooling refrigeration and air conditioning equipment.

Where the municipal code requires that water will not be used once and dumped, but rather must be recycled, the initial cost of a water-cooled installation increases beyond that of a typical air-cooled system and cannot be recouped, by operating economies, in a reasonable length of time.

OPERATING CHARACTERISTICS

The low side pressure of a refrigeration system depends on the application. That is, we can get a good idea of the low side pressure of a walk-in refrigerator and freezer, based strictly upon how it is being used and the refrigerant that is in the system. The high side pressure, however, is more directly related to the temperature of the air passing across an air-cooled condenser. The general rule, that the condensing temperature of the refrigerant is about 30°F higher than the air temperature, will give a fairly good estimate of high side temperature and, by referring to the temperature-pressure chart, of pressure.

With water-cooled equipment, similar reasoning is used, the difference being that the magic number for estimating high side pressure becomes 15 to 20°F. As an example, if the water temperature entering the water-cooled condenser is 85°F, the condensing temperature would be about 100 to 105°F. This corresponds to a high side pressure of about 117 to 126 psig for R-12. For R-22, the pressures would be 195 to 210 psig.

The typical water-cooled condenser is designed for an entering water temperature of 85°F and a leaving temperature of 95°F. A one horsepower refrigeration unit may be required to reject as much as 12,000 Btu/hr. to this water. Under that situation, a water flow of about 1.2 gpm (gallons per minute) is required. You can see how the water used adds up. That's about 72 gallons per hour and 1728 gallons per 24 hours. And that's only for a one horsepower condensing unit. As the installations get bigger, the amount of water also increases. You can see how water conservationists might be a bit on edge as this water drains away into a sewer.

Water temperature tends to vary with the season, whether it's from a city water supply or from a recycled source, such as we'll look at shortly. As the water temperature changes, you would expect the high side pressure to also change, based on the previous discussion. If all we had was a water-cooled condenser and an uncontrolled flow of water, that is exactly what would happen. Actually, water flow is controlled by a water regulating valve.

WATER REGULATING VALVE

The water regulating valve, Figure 10-3, actually regulates the amount of water flowing through a water-cooled condenser in response to changes in high side pressure. As the pressure increases, more water is allowed to enter the condenser. As the pressure decreases, less water is permitted to enter. The net result of this water regulation is that the high side pressure remains at a nearly constant value. While some valves may be factory set, the more common type allows the user to adjust the setting to the desired high side pressure. Typically, R-12 systems are set to hold about 115 to 125 psig, R-22 systems are set for about 200 psig, and R-502 systems are set for about 215 psig.

When water temperature decreases, as in the winter, the efficiency of

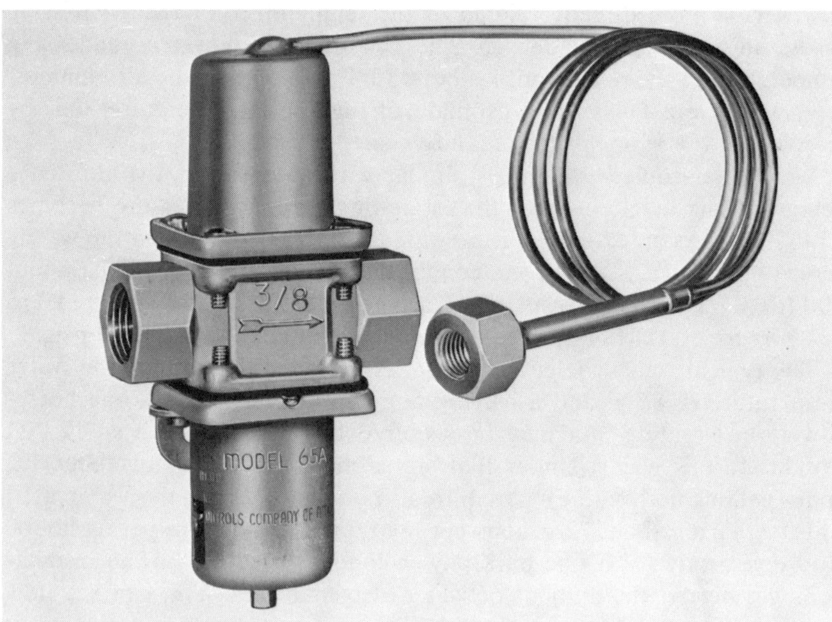

FIG. 10-3. Water regulating valve controls the flow of water through the condenser by sensing head pressure. *Courtesy Controls Div., The Singer Co.*

the condenser increases. The valve will sense this increased efficiency, reflected in a decrease in high side pressure, and will decrease the flow of water. The decreased flow will allow the pressure to increase to the set point once again. If water temperature is high, the valve will allow more of it to flow, and the head pressure will drop down to the set point.

When the system is shut down, the valve should close tight. This makes sense because, when the compressor stops, the high side pressure will decrease and no water will be needed. A fairly common problem is that the valve may not shut off completely, allowing water to flow continually.

PROBLEMS WITH WATER-COOLED CONDENSERS

The purpose of a condenser is to remove heat from the refrigerant, rejecting it to its surroundings. In the case of an air-cooled condenser, the surroundings are air. In the case of a water-cooled condenser, the surroundings are water. If the condenser does not reject heat at its design rate, then its temperature, and therefore the system's head pressure, will increase.

There are three probable causes to be investigated in such a situation.

First, the flow of water through the condenser may be less than required. This is usually caused by a blockage of some sort or a pump that is not working properly.

Second, the temperature of the water entering the condenser may be too high. This is usually due to the cooling tower (more about that shortly) not cooling the water properly. Usually, an investigation of the source of the cooling water will pinpoint the problem.

Third, the heat transfer capacity of the condenser may be impaired by mineral scale, dirt, or algae forming on the water side of the condenser surface. These will act as an insulation, slowing the heat transfer process.

Beyond these factors affecting condenser performance, the next most probable malfunction is a leak. After a period of time, corrosion may cause a leak between the refrigerant and water side of the condenser. A leak in an air-cooled system can be readily detected because the refrigerant escapes into the air where it can be sensed by a leak detector. Refrigerant in a water-cooled system leaks into the water and eventually into the cooling tower or down into a sewer. Unless you are specifically looking for a leak in a condenser, you will not find it. A good practice is to check for traces of refrigerant in the pipe returning water to the cooling tower. If there is refrigerant in the water, you have a leak. Sometimes the water line leaving the condenser must be removed to actually find a trace of a leak.

Since the pressure of the refrigerant is usually greater than the water pressure, most leaks will, at least initially, be from the refrigerant side outward. If the leak is detected early enough, little or no moisture will have gotten into the system and the repair can be a quick one. If the pressures have equalized, there is a good possibility of moisture in the system and an elaborate drying out procedure must be followed after the leak is corrected.

SOURCES OF WATER

The most common source of water for refrigeration equipment is the *city water* supply. Generally, a pressure of at least 20 psig is required at the inlet of the condenser. Use of a water regulating valve is mandatory, so that the city water does not run 24 hours a day.

Cooling towers and closed circuit coolers are growing in importance because of the lower water consumption.

COOLING TOWER

The cooling tower has proved to be quite successful in reducing the amount of water consumed by water-cooled condensers. First, the water passes through the condenser, removing heat from the refrigerant, Figure 10-4. Then, it is pumped to the top of the tower where it is allowed to pass over a filler material, redwood slats and asbestos sheets being commonly used, while air is directed across the flowing water. This movement of the air evaporates a portion of the flowing water. Since the evaporation process is a cooling process, the water that does not evaporate is

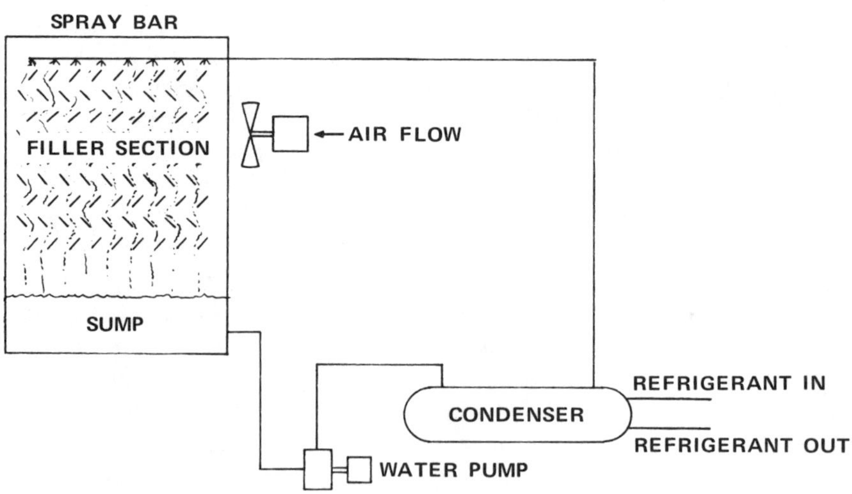

FIG. 10-4. Schematic of cooling tower connected to condenser.

cooled. By the time it reaches the bottom of the tower, the water has been cooled a number of degrees. It is now ready to be recirculated through the water-cooled condenser and to pick up more heat.

During its travels through the tower, about 15% of the water evaporates into the air. The balance of the water is available for recirculation. You can see how this method can save substantial amounts of water. These considerations, however, make the cooling tower less than the boon it appears to be at first glance.

The water, as it passes down through the filler section, not only gets cooled but also cleans the passing air. This air, incidentally, may be

FIG. 10-5. Forced draft cooling towers employ a fan to assure constant air velocity. *Courtesy The Marley Cooling Tower Co.*

driven by a fan, Figure 10-5, or may pass through by natural convection, that is, wind. As it washes the air, it picks up contaminants in the air. In major metropolitan areas, this contamination may be substantial. One component of the air is sulphur dioxide produced by major power companies burning sulphur bearing fuels. Sulphur dioxide combines with the water and forms sulphurous acid, among other compounds. This acid causes the tower to self-destruct over a period of time. It also passes into the condenser and may cause corrosion of the refrigeration system. Dirt in the air is washed out by the water and it too passes into the refrigeration equipment, leading to blockage of strainers, if they are present, or the water regulating valve, if they are not.

As if this were not enough, there are other problems. The water supplied to the tower is the same water that we drink and, interestingly enough, is not 100% pure water, but rather a mixture of water and *other things*. Among the *other things* are minerals, like calcium and magnesium, that tend to remain behind after 15% of the cooling water evaporates. Even though more water enters the tower, to make up for the portion that evaporates, the minerals stay behind and tend to build up in concentration. These minerals have a tendency to coat anything they

FIG. 10-6. Mineral deposits inside a cooling system tube. *Courtesy Calgon Corp., Sub. of Merck & Co., Inc.*

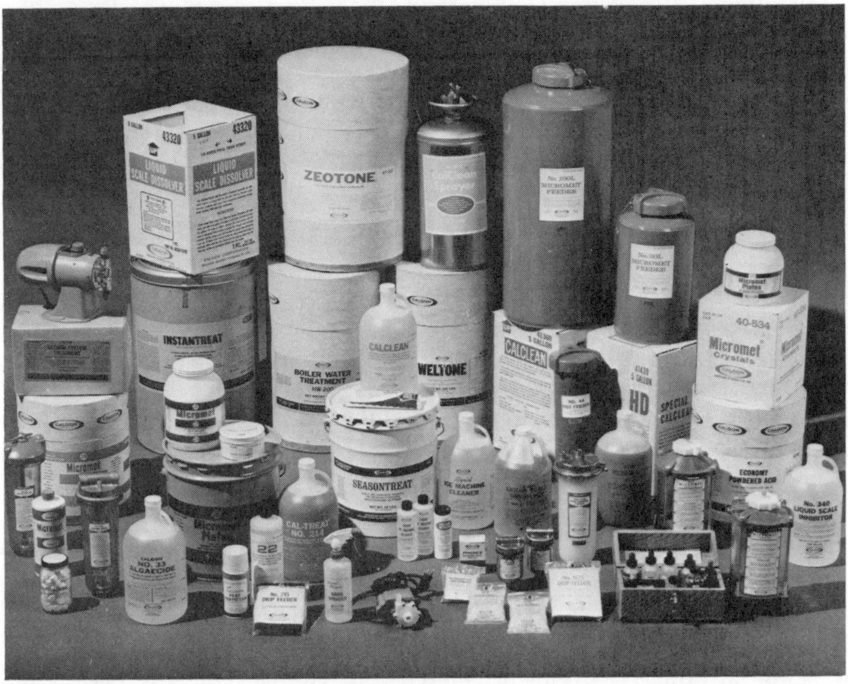

FIG. 10-7. Chemicals used in cooling tower water treatment. *Courtesy Calgon Corp., Sub. of Merck & Co., Inc.*

contact. You can picture the inside surface of a water-cooled condenser, Figure 10-6, after it has become coated with a build-up of minerals. Aside from being ugly, it also has a very low heat transfer rate. It may lose so much efficiency that, no matter how much water is allowed to pass through it, the heat can't be absorbed fast enough by the water. This results in high operating pressures and shortened compressor life.

WATER TREATMENT

At this stage you may be ready to condemn the cooling tower to a watery grave. Actually, cooling towers are quite commonly used, in spite of the problems listed. The answer to any of these problems is good maintenance and regular water treatment. If we treat the water, so that acid does not form and speed the tower on to self destruction, and if we prevent an accumulation of minerals, so that water cooled condensers do not lose their efficiency, then the cooling tower may work out.

Water treatment is a complex subject, beyond the scope of any general type of book. Several things, however, can be said that may be useful to the man in the field. Cleanliness of the water is obviously desireable. Usually, strainers are installed in any water system to remove particles that could jam pumps and control valves.

Beyond this, the most important characteristic of water is the pH number. The pH scale ranges from 1 to 13. Low pH numbers denote water that is said to be *acidic,* while high pH numbers denote water said to be *basic.* Generally, acidic water produces corrosion and basic water encourages the formation of mineral scale. Water at a pH of 7 is said to be *neutral,* being neither corrosive nor scale forming. This is quite simplified, but generally true. Based on this definition, a pH of about 7 would be most desireable, as it is neither acidic nor basic. It turns out that the control of pH number is probably the most important part of water treatment and is usually accomplished by use of chemicals. Once you have determined that water treatment is required, the next step is to consult the experts, usually found with companies that manufacture the chemicals or with firms specifically set up to provide consultation and water treatment service, Figure 10-7.

A technique used to minimize the build-up of minerals in cooling tower water is *constant bleed.* The bleed allows 1 to 2% of the make-up water circulating in the system to drain off into a sewer. In this way, mineral laden water is continually being removed from the system, as purer water is being introduced. Strangely enough, this simple device is very commonly overlooked, yet is very effective in controlling scale formation. Consult the manufacturer's instructions for the amount of bleed to be used.

A final word on water treatment pertains to algae, a group of simple aquatic plants, that can cause blockages in pumps and control valves. Algae grow in cooling towers, usually as a result of the action of the sun on the water. There are chemicals available to kill algae, once it has formed, and to prevent its regrowth.

You can see that cooling towers are not to be installed and forgotten. If any substantial life expectancy is to be realized, and if the refrigeration system is to give long, trouble-free service, a water treatment program must be established. Sad to say, only owners of larger equipment seem to be aware of such programs. The owner down the street doesn't even know such programs exist and usually is faced with service calls to clean out clogged condensers, repair corroded areas of the tower, remove algae build-up, and related problems.

COOLING TOWER SERVICE

A very common problem with cooling tower systems is a water-cooled condenser with excessive scale build-up. This is characterized by con-

FIG. 10-8. Water-cooled condensers. The top unit is a shell and coil condenser typically found in ice makers, water coolers and reach-in boxes. The middle unit is a shell and tube type and the lower unit is a tube-in-tube type. Both of these latter units can be conveniently cleaned with a wire brush after unbolting the removable end plates. *Courtesy Standard Refrigeration Co.*

tinual shutdowns on high pressure control during warm weather. There are two ways to clean this condenser.

If the condenser has removable end plates, Figure 10-8, giving access to the water tubes, the tubes can be cleaned with wire brushes. The tubes can be brushed by hand or an adaptor can be used to attach the wire brush to an electric drill motor. The loose scale is then flushed away, leaving a clean surface once again.

The second technique uses acids to dissolve the mineral build-up. This is quite effective. However, some precautions must be taken. It is particularly important to flush the system after the acid cleaning is complete. If acid is allowed to remain in the system, corrosion could occur, leading to a severe breakdown at a later, but not too much later, date. A test kit that enables you to measure the pH of the water is available. Remember, a pH of about 6.5 to 7.0 is most desirable.

Incidentally, a word about the result of your cleaning activity is in order. On older units, the only thing keeping the water from leaking into the refrigerant may be the built-up scale. Removing this scale may produce higher operating efficiencies, but it also opens a leak between the water side and the refrigerant side. This is not uncommon when acid

FIG. 10-9. Closed circuit cooler resembles a forced draft cooling tower. *Courtesy The Marley Cooling Tower Co.*

solutions are used since you really can't see the condition of the condenser tubes. It might not be a bad idea to prepare the customer for a major problem before you start to clean the condenser.

The body of the tower itself may begin to corrode. Usually, good cleaning and application of a rust inhibiter will add some years to the unit. Repairs of completely corroded areas, with sheet-metal patches and epoxy compounds, has also helped delay the inevitable. The corrosive action is usually magnified by an accumulation of scale and dirt in the tower sump. Yearly cleaning of the sump is a worthwhile exercise.

THE CLOSED CIRCUIT COOLER

Since the water washing over the filler in the tower also passes through the water-cooled condenser, many of the contaminants in the air will filter into the condenser. Not only is the tower threatened, but the refrigeration system itself is threatened. The closed circuit cooler, Figure 10-9, is a good solution to this problem. Like all good solutions, it costs money. In fact, a good closed circuit cooler may be as much as twice as expensive as a cooling tower.

Two water circuits are used in the closed circuit cooler. One circuit pumps sump water to the top of the cooler. From there, the water trickles

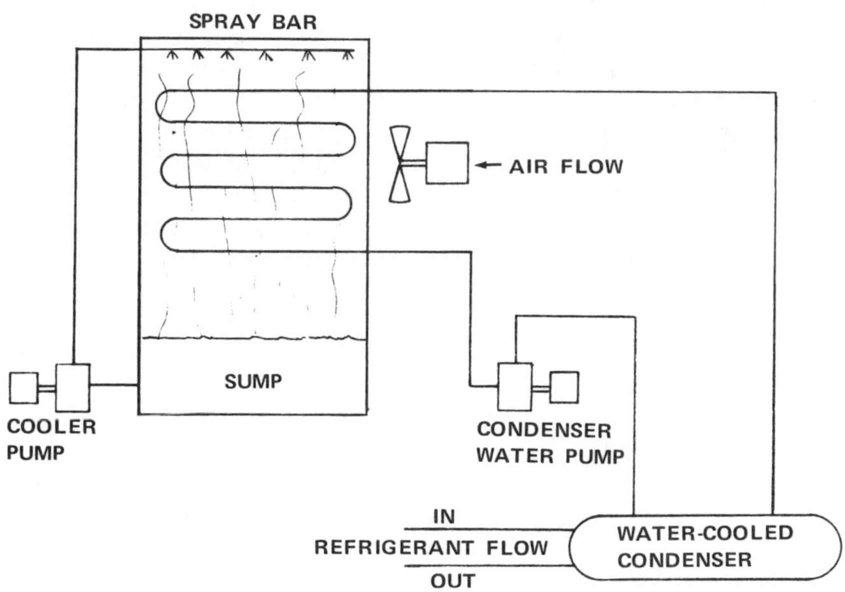

FIG. 10-10. In the closed circuit cooler, condenser water is cooled without contacting the outside air.

down over the tubes, Figure 10-10. The tubes in the second circuit contain water from the condenser. Water trickling over the tubes evaporates and cools the water within the tubes. The water within the tubes is never in contact with the air and therefore never becomes contaminated. Since the condenser is never subject to acidic water nor to water with high mineral content, most of the condenser problems are eliminated. The cooler, of course, is subjected to all the problems of a typical cooling tower and must be maintained in a similar manner.

WINTER OPERATION

Refrigeration is a year round requirement. Therefore, both the cooling tower and closed circuit cooler must be able to provide cooled water all year. The water regulating valve does a good job of maintaining a predetermined high side pressure, no matter how low the water temperature drops. However, on very cold days, water velocity may possibly be reduced to the point where the water can freeze in the pipes. This is a serious problem because the pressure generated by freezing water will rupture the water lines. Several techniques have been developed to prevent freeze-ups in cooling towers and closed coolers.

One technique is to have a temperature control cycling the tower fan. If the water gets too cold, indicating that not much of it is needed to cool the condenser, the fan will shut off, allowing heat to build up in the tower. Another technique, used with some success, is to install heaters in the sump, again controlled by an outdoor thermostat that turns them on whenever the outdoor temperature drops too low. A third way to counter the effect of winter weather is to add antifreeze to the tower water. Finally, the addition of a false load to the condenser water (electric immersion heaters or thermostatically controlled steam coils have been used effectively) will help keep the system sufficiently warm and prevent freezing with its accompanying problems.

11

Supermarkets

The food retailer is one link in the chain that sees food flow from the farm to the consumer. It has been estimated that as much as 50% of the food purchased by consumers requires refrigeration, with about 10% being frozen food. The refrigeration equipment found in the food market has two purposes. It must maintain the food at sufficiently low temperatures so that spoilage does not occur. It must also present the food in a way that inspires the consumer to purchase. In short, the two functions can be expressed as *preservation* and *merchandising.*

The equipment already described falls into general categories known as reach-in and walk-in refrigerators and freezers. There is little to be added to that, as we look at supermarket equipment. There are some variations to be sure, but they are minor and should pose no problem.

The broad categories of refrigeration encountered are:

Type	Temp.
Produce refrigeration	35-40°F
Meat refrigeration	
Display	33-36°F
Storage	28-32°F
Dairy Refrigeration	35-40°F
Freezing	0°F
Ice cream storage	—15°F

You can pretty well classify anything found in supermarket refrigeration under one of these broad categories. Manufacturers' tables are available showing more precise temperatures for specific products but, practically speaking, being able to hold air temperatures in the ranges shown will usually be quite acceptable.

A good number of supermarkets have long-term refrigerated storage facilities. This is particularly true of stores that are somewhat remote from central storage areas and do not receive deliveries every day. Typically, the long-term storage, in the form of walk-in refrigerators and freezers, will be the equivalent, in holding capacity, of the display equipment on the selling floor. In any case, they are similar in all respects to the walk-in equipment previously described. Most of this chapter will be devoted to describing display type equipment and the refrigeration systems they incorporate.

CONFIGURATIONS

Display refrigerators and freezers generally fall into two categories: vertical and chest type. The vertical display box has either sliding or swinging glass doors that permit the contents of the box to be seen from a fairly long distance, Figure 11-1. From time to time, fashion changes and a vertical box may be popular for dairy products one year and not popular the next. However, no hard and fast rules can be set about what products will be displayed in which type of display case. This is pretty much a store layout and merchandising decision, rather than a refrigeration requirement. The vertical boxes have been described before, in the discussion on reach-in refrigerators and freezers, and there is little more to say.

FIG. 11-1. Vertical display box. Louvers at the bottom right indicate the location of the condensing unit. *Courtesy La Crosse Cooler Co.*

The chest type display cases, however, are unique to food markets and further discussion is warranted. The true chest type display case is usually found as a freezer, Figure 11-2. It takes advantage of the fact that cold air is heavier than warm air and that the cold refrigerated air will sit in

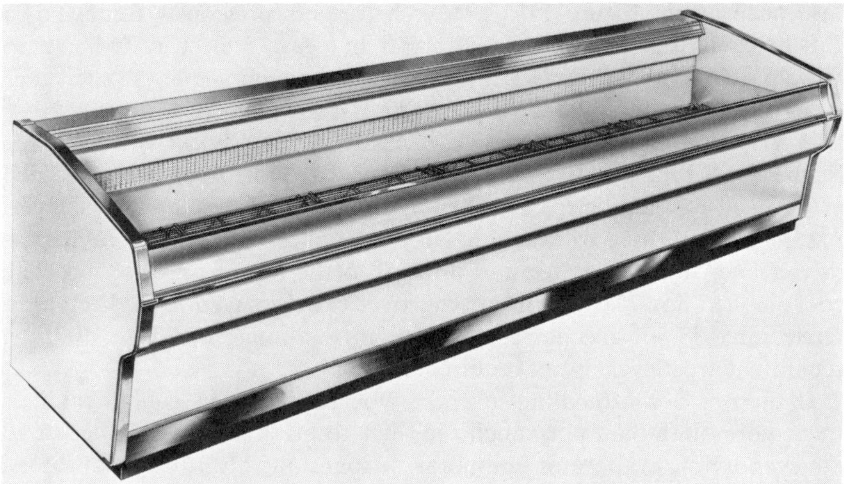

FIG. 11-2. Chest type display case is available as a self-contained unit. More frequently, it is connected in a row to form a frozen food aisle. *Courtesy Tyler Refrigeration Corp.*

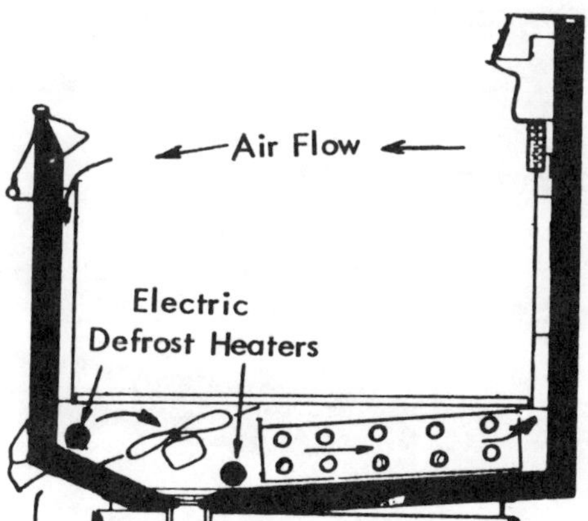

FIG. 11-3. Air flow pattern of a chest type display case. *Courtesy Tyler Refrigeration Corp.*

a case with relatively little spillover onto the floor. For this reason, no doors need be used. Of course, this is desirable from the marketing point of view and probably more of this type of freezer is used in supermarkets than any other. The evaporator is located at the bottom of the case and fans are used to pull air from the case, cool it, and send it back into the case once again, Figure 11-3. As with freezers previously discussed, a defrost cycle is usually provided, either hot gas or electric. Other than the physical configuration of the cabinet, it is quite similar to the other types of refrigerators already described.

A variation of the true chest type has a vertical back with shelving, Figure 11-4. It is usually used as a refrigerator. Some items are contained in the well of the chest and other items are displayed on the shelves. Again, fans are used to draw air out of the chest and, in this case, pass the air over the evaporator and upward to the top of the shelving. The cool air then flows, as an air curtain, over the displayed food, keeping it refrigerated. Since this is refrigeration in the range of 35 to 40°F, a separate defrost cycle is not required.

If there is a shortcoming in chest type refrigeration equipment it is that, more often than not, much food has to be shifted to gain access to the evaporator, evaporator fan motors, expansion valves, defrost heaters,

FIG. 11-4. This case combines the advantages of chest type construction with the high visibility of a vertical display case. *Courtesy Tyler Refrigeration Corp.*

etc. A simple operation, like checking for refrigerant leaks, takes on the aspect of a major project if the case has not been emptied of its contents. Once access has been gained, troubleshooting and system analysis is quite similar to that already described.

MEAT STORAGE AND DISPLAY

A characteristic of refrigeration is that, during the cooling process, air passing over the evaporator is both cooled and dehumidified. As a result of this dehumidification, the air in the refrigerator tends to become quite dry and, in turn, the contents of the box also tend to dry out. When the box holds packaged goods, this poses little problem. When the contents are meat or produce, however, dehydration can occur. Dehydration makes the product less appealing in appearance, hence more difficult to sell, and reduces the length of time before spoilage.

It is a rather common procedure to spray vegetables with water each morning, to replenish any moisture that refrigeration has removed. This procedure is not possible with meats. Several techniques have been applied, over the years, to reduce the amount of moisture removed from meat during storage.

EVAPORATOR PRESSURE REGULATOR

If the surface temperature of an evaporator is below the *dew point* temperature of the air, condensation will form on its surface. The *dew point* of air is defined as the temperature at which water vapor in the air will condense. Since meat storage temperatures are in the low thirties or high twenties, it is obvious that refrigerant temperatures in the evaporator will be on the order of 10 to 15°F. These temperatures are well below the dew point temperature of air in most places.

The further the coil temperature is below the dew point temperature, the more moisture will be extracted from the meat.

The purpose of an evaporator pressure regulator, Figure 11-5, is to limit the minimum temperature of the evaporator coil. The evaporator pressure regulator is installed in the suction line close to the evaporator. It is designed to maintain a fixed pressure on the upstream, or evaporator, side of the valve. With a fixed pressure setting, the temperature of the refrigerant in the evaporator is also fixed at a minimum point.

The effect of this valve is to reduce the rate of evaporation of moisture from the surface of meat being stored. Even though suction pressure at the compressor might be quite low, due to a low load on the system, the pressure in the evaporator will be relatively high. As an example: an evaporator pressure of 23 psi might be set on the valve. The pressure at the compressor might drop to the low teens, but the pressure in the evap-

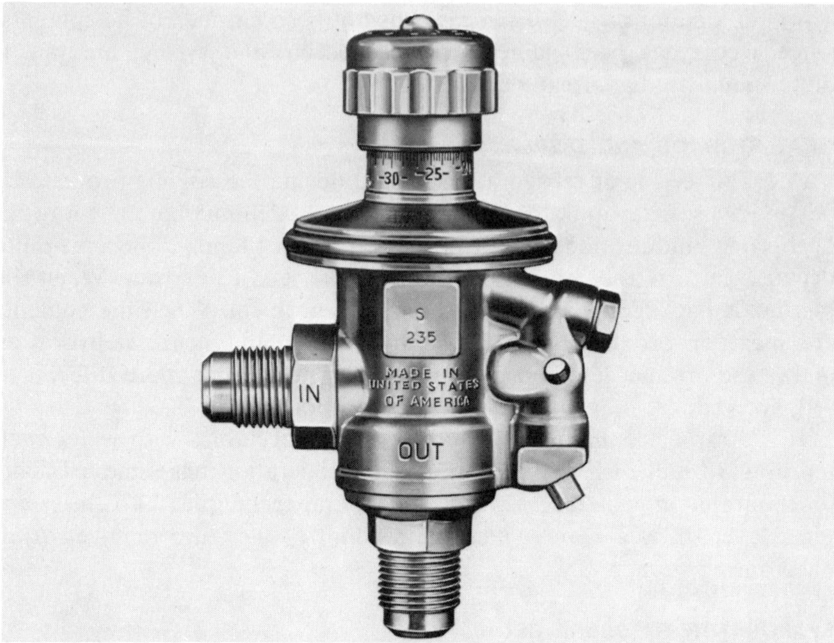

FIG. 11-5. Evaporator pressure regulator can be adjusted to any desired pressure. A port into the inlet side of the valve allows the pressure to be checked by a gauge. *Courtesy Controls Div., The Singer Co.*

orator, and therefore the temperature, will remain at the higher setting. All dehydration is not eliminated, it can never be, but it is substantially reduced.

GRAVITY COILS

Low air velocity in a walk-in or display meat box means less dehydration. The movement of air over the surface of the meat removes moisture from its surface in much the same way as air flow over our bodies causes the evaporation of perspiration. If the velocity of the air in the box can be reduced to an absolute minimum, and the box still be kept cool, then meat storage can be prolonged.

The gravity coil, Figure 11-6, does exactly this. No fans are used in such coils. Convection, as a result of cool air falling and warm air rising, produces the necessary air motion. Gravity coils are characterized by large surface areas compared to the blower coil type of evaporator, Figure 11-7. They are ceiling mounted and positioned to provide adequate room for natural convection to occur.

FIG. 11-6. Gravity type coils in the top of this meat display case minimize surface dehydration and meat discoloration. *Courtesy Tyler Refrigeration Corp.*

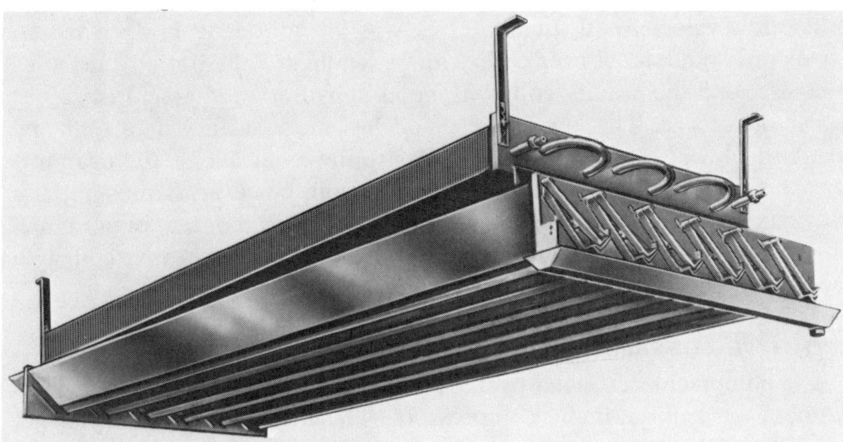

FIG. 11-7. Large, ceiling-hung gravity coil commonly found in walk-in meat and florist cases. *Courtesy Kramer Trenton Co.*

On the positive side, they are quiet, use less energy, since fans and motors are not used, and reduce dehydration to a minimum. On the negative side, they are, in general, more expensive than blower coils of comparable capacity and they are much bigger and bulkier than the blower coil, requiring a higher ceiling for installation. From the standpoint of

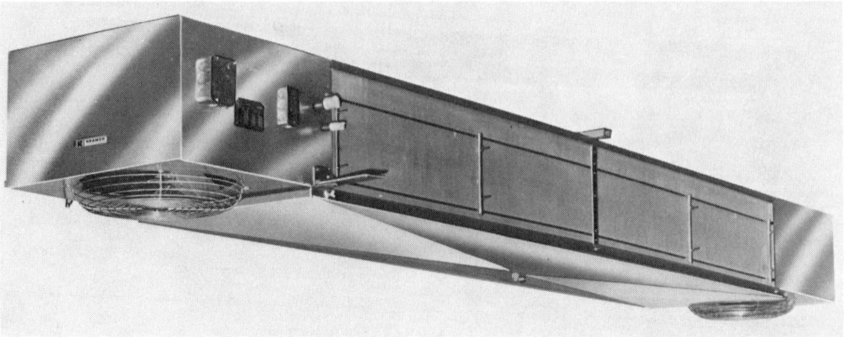

FIG. 11-8. Low velocity coil combines the low air velocity of the gravity coil with the compact size of the blower coil. *Courtesy Kramer Trenton Co.*

service, they compare favorably with the blower coil since, having no fans and motors, there is less to malfunction. Servicing the refrigeration system is the same as for the systems previously discussed.

LOW VELOCITY COILS

The low velocity coil, Figure 11-8, was introduced to bridge the gap between a standard blower coil, with rather high dehydration but compact size, and the gravity coil, with good dehydration characteristics but rather bulky size. The low velocity coil has more surface area than the standard blower coil, but does use fans to move air across the evaporator. The amount of air passing through the coil, however, is substantially less than with the standard coil. Airflows that are reduced by as much as 1/3 are commonly found in such coils. The net result is a coil with acceptable size and dehydration characteristics.

OPEN TYPE COMPRESSORS

The equipment described to this point was assumed to have either full hermetic or semi-hermetic compressors. The fact is, that once you get beyond the small reach-in boxes and small walk-ins, you begin to see more and more open type compressors in use, Figure 11-9. There is a continuing battle between a steadily decreasing number of service managers who are pro-open compressor and those who are pro-hermetic. It is a battle that probably will never be resolved to everyone's satisfaction and only time will eventually tell who is right.

Those who like open compressors point to the numerous units that are still in service after twenty or more years of reliable operation. They point out that the open compressor does not rely on suction gas cooling

FIG. 11-9. Open compressors are available in a variety of sizes. With the appropriate pulley and motor, any one of them can be adapted to a number of applications. *Courtesy Lehigh Inc.*

for efficient operation, therefore a refrigerant leak or a defective TXV need not lead to a total compressor failure. The point out that, if a motor should burn out, it can be readily replaced without having to open the refrigeration system, Figure 11-10. They point out that standby motors can be kept on hand much more easily, meaning inexpensively, than standby compressors. They point out that an open compressor can be stripped down, and any defective parts readily replaced, by a trained serviceman while a similar problem in a hermetic compressor would necessitate replacing the entire compressor.

Those who like hermetics acknowledge much of what the pro-open people say, however, they also indicate that the key to long life for such

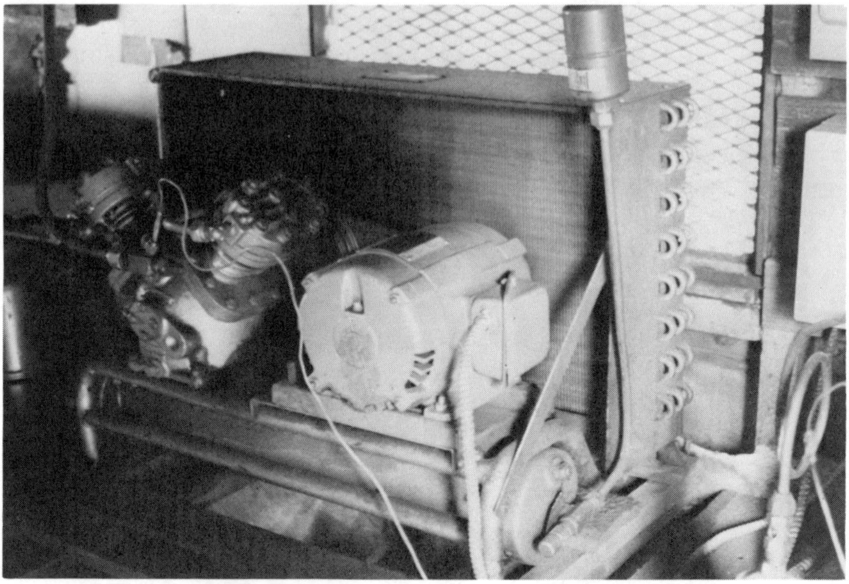

FIG. 11-10. Open type compressor in a typical equipment room. The motor can be changed without opening the refrigeration system.

equipment is a good, reliable service person. There are, in fact, many good, reliable service people in this country but, as often as not, they are running a service business and sometimes have to send out a man who may not be quite as good and reliable, to do the actual work. The hermetic people admit that all they need are parts changers, for most of their repairs, and they want it that way. A parts changer can get the equipment back in operation quickly, if he has the parts. Hermetic compressors need fewer repair parts, mainly because they aren't repaired in the field, but are usually changed if they malfunction. The skill required to change a compressor is much less than that needed to change valves or piston rods, or oil pumps or unloaders, or any of the many other components encountered in open compressors. The extra service that might be needed is offset in large part by the lower initial cost of the equipment. While many would argue about it, the sad fact is that initial cost, rather than long range reliability, is increasingly the main consideration in equipment selection.

MULTIPLEXING

In the early days of display refrigeration, the size of food stores was relatively small. A typical store might have one or two refrigerators, and possibly a freezer, and that was about it. Each refrigerated display case had its condensing unit built in or, if remote, had its condensing unit in the basement or storeroom. In more recent times, the small store has pretty much been replaced by the supermarket as the main source of family groceries. The large food markets have, typically, a dozen or more of both refrigerator display cases and freezer units.

Rather than use self-contained boxes in such applications, it is advantageous to use remote systems, the main reason being that more storage volume is available in a box with a remote condensing unit. An additional consideration has recently come to light. The heat from the condenser, when added to the cooling load of the store, increased the requirement for air conditioning. By remoting the condensing units, no additional heat is added to the store. Still another aspect of this energy conservation viewpoint has been considered. The spillover from the display freezers and refrigerators can be drawn away from the refrigerated food section, by floor ducts, and introduced into the air conditioning system, still further reducing the cooling equipment requirement. In line with this thinking, the heat of condensation, if available at a central location, could be added to the heating system in the winter, to reduce heating system requirements. In the years when energy and water were

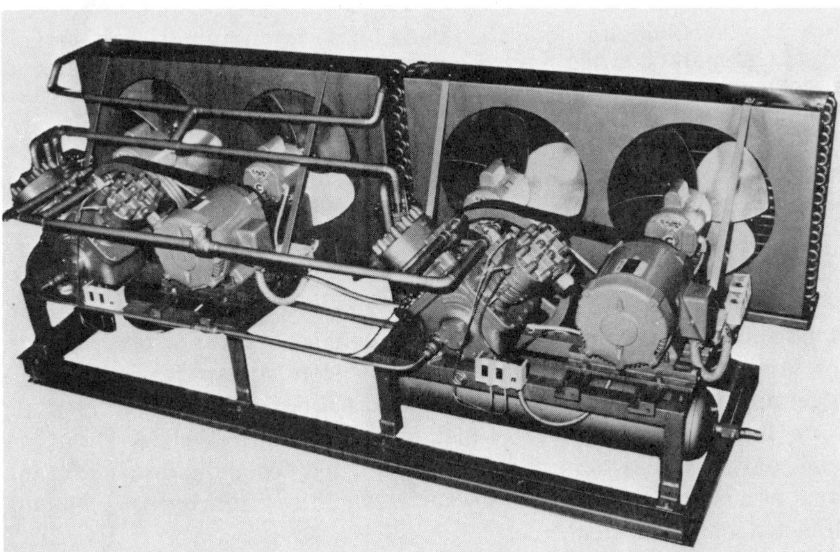

FIG. 11-11. These tandem condensing units serve a number of refrigeration cases. If one unit is down for repairs, the other can sustain the load for extended periods of time. *Courtesy Lehigh Inc.*

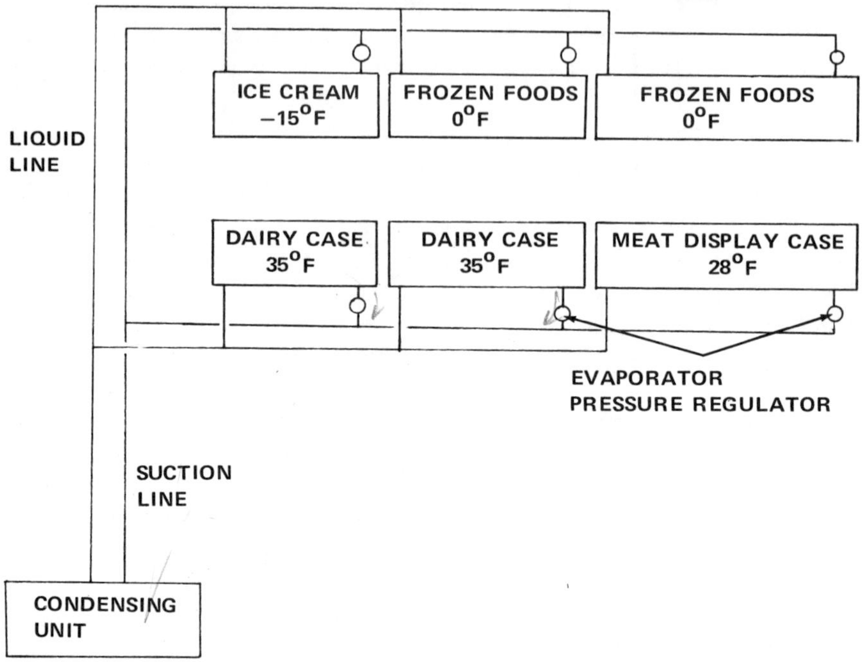

FIG. 11-12. Multiplexing system used in the refrigerated foods section of a small market. While the ice cream case could have been connected directly to the compressor, it is equipped with a evaporator pressure regulator in the event of an alternate use.

equally taken for granted, no thought, or at least very little thought, was given to such considerations. Now it is the rare store layout engineer who does not offer such an option in the development of a supermarket store plan.

With all the condensing units in one remote location, it was a natural step to think of multiplexing, that is, to connect several display cases to the same condensing unit. It is now not uncommon to have all the refrigeration equipment in a store connected to one or two large condensing units, Figure 11-11. Instead of maintaining two dozen refrigeration systems, only one or two must now be maintained. To avoid disaster, in the event of an equipment failure, a second, standby compressor is commonly designed into the system.

Even though, as we saw earlier, there is a range of storage temperatures that must be maintained, we can maintain them with multiplexing. Each unit connected to a compressor will have its own evaporator pressure regulator, Figure 11-12. The evaporator pressure, hence tempera-

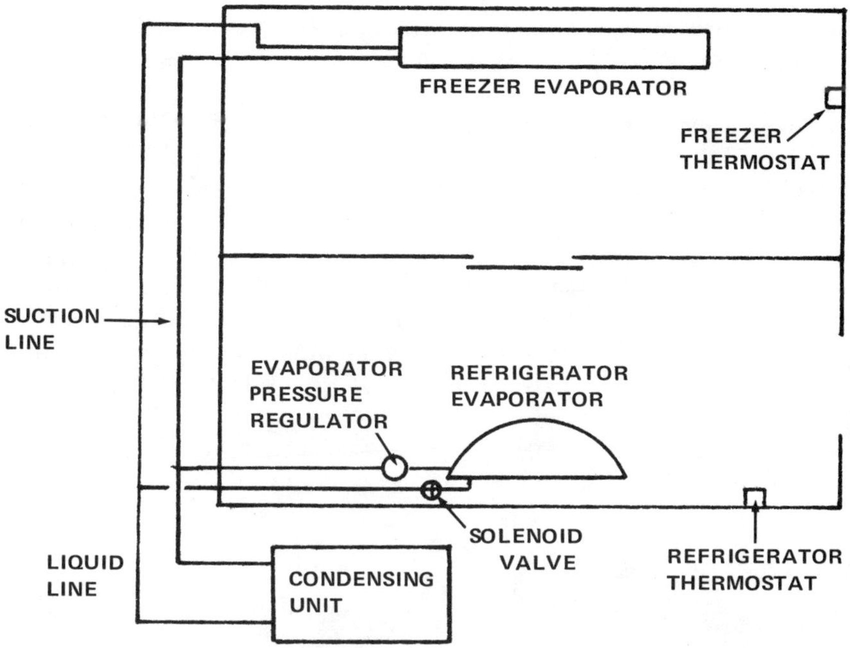

FIG. 11-13. This two-temperature walk-in box can be found with two refrigeration systems or with a single condensing unit, as shown.

ture for each evaporator, can be set according to its application. Meat storage boxes can be set to maintain a 23 psig evaporator pressure, while dairy cases can be set for 12 to 15 psig, with frozen food cases set at 0 to 5 psig. Ice cream storage boxes can be directly connected to the compressor, since they are the coldest boxes, and the compressor must run at a suction pressure low enough to keep them cold. We would, therefore, see the temperatures listed at the beginning of the chapter being held in each case because of the action of the evaporator pressure regulator.

This sounds a bit easier than it actually is. The manufacturer of such systems has guidelines for the technician to follow in connecting the various types of boxes. It is not something to be done indiscriminately by anyone and certainly not without the advice of the equipment manufacturer.

TWO-TEMPERATURE WALK-IN BOX

Prior to multiplexing, it was not at all uncommon to see a walk-in box divided into two compartments. Typically, the outer door would open into a refrigerator section. Within this section was a second door that opened into a freezer compartment. While many of these configurations had two completely independent refrigerations systems, one medium temperature and one low temperature, in many cases one system did it all.

The low temperature evaporator was directly connected to the compressor, Figure 11-13. The medium temperature evaporator had an evaporator pressure regulator, connected in the way previously described. It was, therefore, possible to have two distinct operating temperatures, using one compressor. In addition to the evaporator pressure regulator, however, there were additional temperature controls in both the refrigerator and freezer sections. A wall-mounted thermostat was connected to a solenoid valve, Figure 11-14, in the liquid line supplying the refrigerator coil TXV. When the thermostat was satisfied, the solenoid valve closed, stopping the flow of refrigerant to the evaporator. The low temperature circuit would remain open and refrigerant would continue to flow. When the low temperature box was cold enough, as determined by

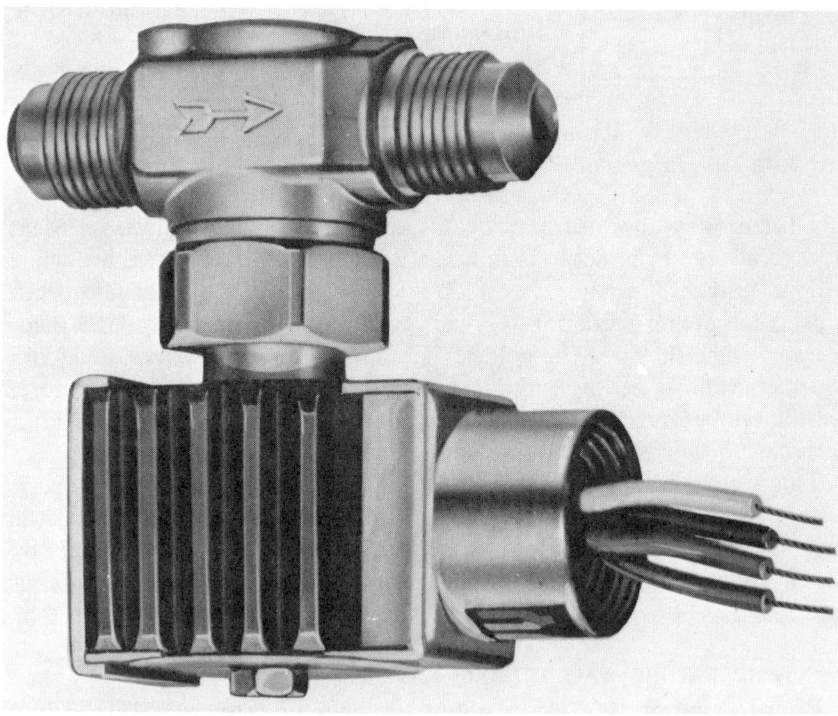

FIG. 11-14. Solenoid valve used in two-temperature applications can be wired for either 115 or 230v applications. Note the arrow indicating direction of flow. *Courtesy Sporlan Valve Co.*

the freezer thermostat, then the compressor would shut down. You can see that this is quite similar to multiplexing and was really the first step that eventually led to the present supermarket systems.

PIPING

The refrigerant lines used in supermarket installations very often have to run over a hundred feet from the display floor to the condensing unit. Growing use is being made of roof mounted condensing units for refrigeration systems. This has been a practice in air conditioning for a number of years, but only recently has it gained popularity in refrigeration installations. In some cases, the compressor will be located in an equipment room and a remote air-cooled condenser will be on the roof. In other situations, both the compressor and condenser will be located on the roof. In either case, you have a situation where a line is carrying refrigerant vapor in a vertical direction.

The refrigerant oils used today generally are quite miscible, or easily mixed, with R-12, R-22 and R-502 refrigerants. Liquid oil mixed with liquid refrigerant poses no problem, as far as it's being able to move

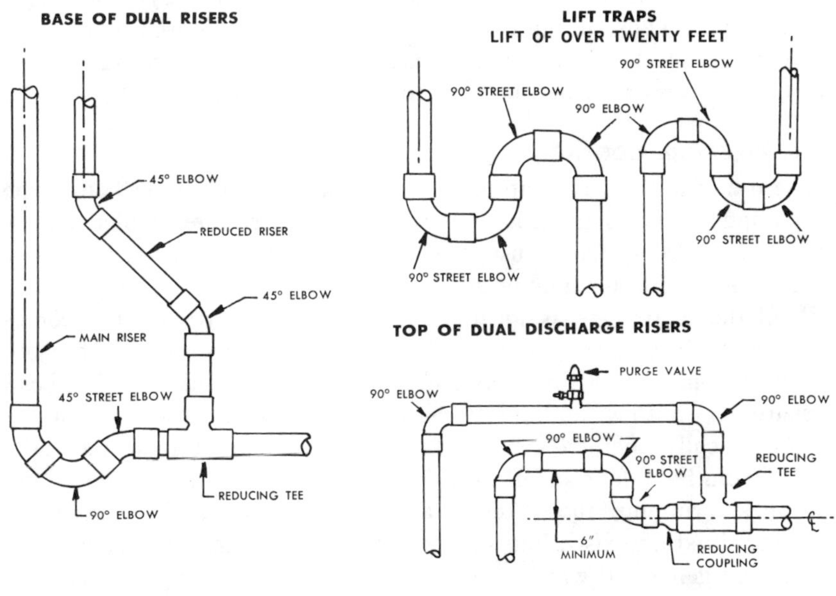

FIG. 11-15. Dual risers, traps every twenty feet of vertical lift, and proper connections at the top of the dual riser are all needed to assure adequate oil return to the compressor. Courtesy Dunham-Bush, Inc.

through a refrigerant line is concerned. The problems arise when the liquid oil is moving through the discharge line or the suction line where the refrigerant is in the form of a vapor. The only thing that carries the oil along is the velocity of the refrigerant. This problem was mentioned earlier, with regard to sizing lines. The problem in multiplexed systems is even more acute.

As the load on the refrigerant system is reduced, the compressor tends to unload and the velocity of the refrigerant vapor will decrease. This decreased velocity may very well not be great enough to carry the oil through the system. A solution to this problem is found in the use of double raisers on the discharge side and/or suction side of the system, Figure 11-15. When refrigerant is to flow upward, two risers are installed side by side. One line is sized to carry the refrigerant flow under full load conditions. The second riser is sized smaller to enable the refrigerant, under low load conditions, to maintain its velocity.

A trap is installed at the bottom of the larger riser to collect oil under low load conditions. This oil will accumulate until it completely blocks the flow of refrigerant through the larger of the two lines. The refrigerant must now flow through the only remaining line, namely, the one with the reduced diameter. The velocity of the refrigerant through this smaller diameter tube will be about the same as full flow through the larger diameter tube. Thus, all but the small amount of oil captured by the trap will flow through the system and eventually return to the crankcase. Remember: oil anywhere but in the crankcase is useless.

HEAD PRESSURE CONTROL

The use of a water regulating valve has already been described as the most popular means of head pressure control for water-cooled equipment. Air-cooled systems must also have some measure of head pressure control and several techniques are in use today.

If all the equipment is located in an equipment room, it becomes necessary to provide some means of ventilation for this room. The problem, in this instance, is to remove heat from the condenser so that head pressure does not get too high. Perhaps the most widely used system is a thermostatically operated exhaust fan. As the temperature in the room becomes too high, let's say above 95°F, the fan automatically starts and continues to run until the air temperature drops to about 85°F. In winter, the same situation exists, but the fan has to run less frequently, as much of the heat generated will be lost due to low outdoor temperatures.

Outdoor equipment is usually provided with a device that will control low pressure rather than high pressure. As the outdoor temperature falls, the high side pressure also falls. The TXV will pass an adequate amount

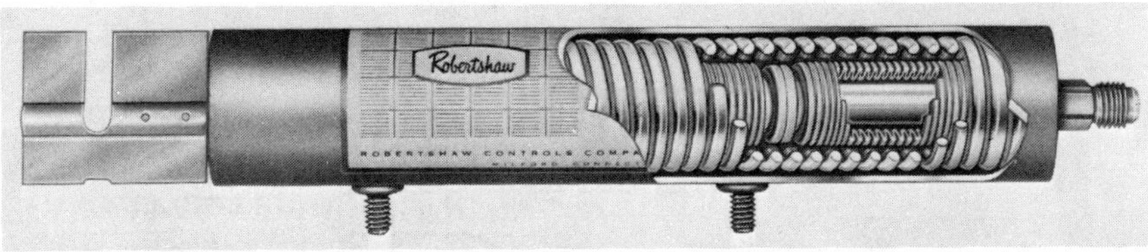

FIG. 11-16. As high side pressure increases, it actuates this damper operator and opens the damper to increase air flow. With a decrease in pressure, the damper closes to restrict air flow. *Courtesy Robertshaw Controls Co.*

of refrigerant as long as the high side pressure is perhaps 75 to 80 psig higher than the low side. Actually, the rating of TXV's is based on a 100 psig pressure differential through the valves but, in most applications, it will work quite well with somewhat lower differentials. In any case, as the outside temperature drops, the high side pressure may drop to unacceptably low levels unless some control is exerted. This control has taken two popular forms.

One is the use of dampers covering the face of the condenser coil. The dampers are operated by an actuator connected to the high side pressure, Figure 11-16. As the pressure decreases, the dampers close, restricting the flow of air through the condenser coil. As the airflow drops, the temperature, hence pressure, of the refrigerant increases. In this way, control can be achieved over a wide range of outdoor temperature conditions. As air temperature increases, pressure will also increase and the dampers will move toward their full open position, to allow maximum airflow.

A second technique also blocks off part of the condenser, making it less efficient. While outside surface was reduced by the face damper control, inside surface is reduced in this technique. Liquid refrigerant is forced to back up in the condenser, reducing the amount of active surface available for actually condensing refrigerant. One method used, Figure 11-17, uses two valves to do the job. The inlet pressure regulator senses the pressure of the refrigerant in the condenser. As this drops, as a result of low ambient, the regulator closes down, forcing the pressure to build up at the bypass valve. This rising pressure forces the bypass valve open, permitting high temperature refrigerant to enter the receiver. This refrigerant pressure causes the check valve to close, stopping flow to the receiver from the condenser. Liquid then accumulates in the condenser, reducing the condenser surface and gradually causing pressure in the condenser to rise. The inlet pressure regulator and the bypass valve are working together to maintain a predetermined high side pressure as ambient temperature falls.

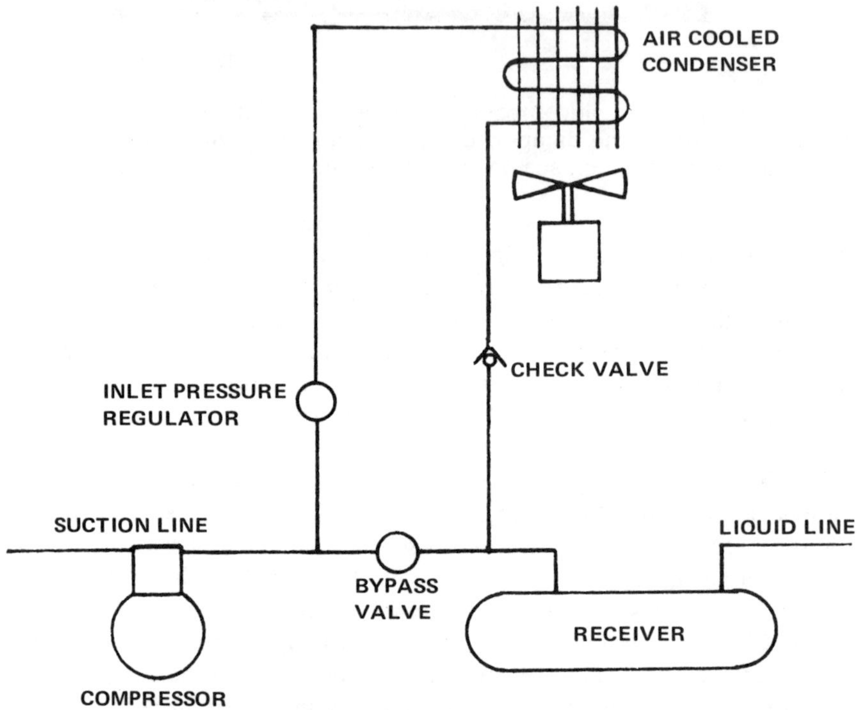

FIG. 11-17. A by-pass system used to maintain sufficient head pressure while the condenser is operating in low ambient conditions.

This technique is quite effective and is enjoying increasing popularity as the number of outdoor refrigeration condensing units increases. One characteristic of such a system is that it usually employs much more refrigerant than a typical system with no such control. As an adjunct to this system, the condenser fan may be wired to a reverse-acting pressure control that will cause it to remain off until condensing pressure builds up to a preset value. This is quite commonly the first step in head pressure control with this method.

12

Food Service

Food Service is a catchall phrase for the activities surrounding the preparation and serving of food. Refrigeration equipment in the food service category is generally designed for short-term storage and for display of prepared foods at the point of purchase or final preparation, Figure 12-1. The display cases used to tempt you into buying that banana cream pie that you know you shouldn't have, the sandwich table containing the tuna fish, egg salad, chicken salad, etc., that will eventually be made into your lunch, the counter top soda machine, and the cold plate containing eye-pleasing salads are all examples of food service equipment.

The major difference, in all of these applications, is the form of the evaporator. It might be a flat plate, a blower coil, a coil of tubing immersed in water, a cylindrical shell with a tube bonded onto its outer surface, or perhaps a sandwich of embossed aluminum plates of the

FIG. 12-1. Food service line equipped with a ice flaker/dispenser, a soda system and a bulk milk dispenser.

type commonly found in domestic refrigerators. Other than the evaporator, there are few mysteries to be found in this type of equipment.

WORKING EQUIPMENT

The *sandwich table* has a storage compartment containing a blower coil, as well as a number of small trays containing sandwich *fixins,* e.g., tuna fish salad, tomatoes, mayonnaise, pickles, egg salad, sandwich meats, etc. The top of the sandwich table is not solid, but is rather made up of a grid supporting these trays, Figure 12-2. They protrude downward into the cooled interior of the box and their contents are therefore kept at a suitable temperature. Capillary type units are widely used in this application, in conjunction with thermostats mounted on the face of the evaporator coil.

A *dough retarder* is a low refrigerator, with a counter surface used for working dough, commonly found in a pizza restaurant. The purpose of the refrigerator is to retard the action of yeast, so that dough can be mixed in large batches and used in small portions over a period of time. Actually, any refrigerator used for this purpose is called a dough retarder. A real problem, in talking to someone over the phone about a dough retarder, is to get an accurate description of what the customer means by this term. Any refrigerator used to store dough is, in effect, a retarder.

FIG. 12-2. This sandwich table has a self-contained refrigeration system with the condensing unit behind the louvered panel. *Courtesy Bastian-Blessing Co., Div. of Astro Controls Inc.*

Again, either thermostatic or capillary tube systems can be found in this application.

The old standby, the reach-in box, can be categorized as food service equipment. There are a number of variations on the standard box that make it more effective in this application. A common variation is the *pass-thru* box. This box has doors at both the front and rear of the unit. Typically, the box is loaded through the rear, by those who prepare the food in the kitchen, and is unloaded at the front, by those who sell the food on the service line. Other than cabinet configuration, there is nothing else unique to this type of unit.

A *roll-in* box is another type of standard reach-in. The bottom of the box either has a low ramp or is installed flush with the floor, so that a cart can be rolled into the box, Figure 12-3. The cooks will prepare a cart containing a number of trays of food. They will roll the cart into the box and it will remain there until needed by the people on the food service line. Both the pass-thru and the roll-in are designed to smooth traffic flow through the kitchen and service area and to minimize food handling and confusion.

The refrigerator/freezer combination is quite common around the

FIG. 12-3. Roll-in refrigerator has tracks in the floor of the box to permit easy movement of the roll-around racks. The condensing unit is at the top of the box. Mullion type evaporators are located between the doors. *Courtesy Foster Refrigerator Corp.*

home. In commercial applications, it is decidedly less common. When such a combination exists, it is almost always provided with two completely independent refrigeration systems. The typical domestic refrigerator/freezer will not be opened and closed as many times in a week as its commercial counterpart will be in a day. The commercial unit, therefore, must have a heftier refrigeration system, as well as more frequent defrost cycles. Over the years, it has been found that two refrigeration systems give better control and longer life than the single system found in the home unit. While it is not uncommon to find domestic boxes applied to a commercial food service operation, the large majority of combination units are the commercial type. In servicing such a unit, there should be no difficulty as long as you remember that there are two systems, completely independent of one another, and they can be treated as such.

DISPLAY EQUIPMENT

The purpose of display equipment is to keep food cold and, at the same time, display it to potential customers in such a way as to motivate them to buy, Figure 12-4. The reach-in refrigerator, equipped with glass

FIG. 12-4. A reach-in display case. The space occupied by the condensing unit is not important since low shelves are not popular with consumers. *Courtesy Foster Refrigeration Corp.*

doors, is the standard display box. The main difference between the display unit and the regular unit is that the display box will usually have a larger refrigeration system. The heat gain through the glass panels, even though the panels are double glass with a sealed-off air space, is greater than the heat gain through a standard door. Usually, the condensing unit is one size larger on the display box.

A cold plate is, as the name would suggest, a plate kept at a cold temperature. Salads, sandwiches, small containers of milk or juice and other small items can be attractively displayed on such a cold surface for an extended period without spoilage. Many cold plates are nothing more than a bare tube evaporator covered by a stainless steel panel. Other cold plates are part of a complete refrigerator, with the top being the cold plate and the bottom being a low, reach-in refrigerator. In this case, two evaporators are connected to the same condensing unit. Each evaporator may have its own control or they may be connected, in series, to a single control.

A dessert case, Figure 12-5, is commonly used to display fruit, pud-

FIG. 12-5. This pass-thru display case will be filled with deserts or salads through the rear door. Upon opening that front door, the customer can make a selection. *Courtesy Bastian-Blessing Co., Div. of Astro Controls Inc.*

dings, pies, cakes, etc. in a manner designed to prompt the impulse buyer, or diner, before he gets to the cashier. The case may range in size from one to four or five shelves, either sitting on top of a serving counter or incorporated in a low, storage refrigerator. The dessert case has glass doors and walls for maximum visibility of the contents and the low refrigerator, if any, will have solid doors. In such an arrangement, each refrigerator will have its own evaporator, usually with its own control. It's not unusual to see several small cases of this sort connected to one condensing unit. Whenever more than one evaporator and flow control is used, the control will be a thermostatic expansion valve. The capillary tube is not employed in these applications because of its operating characteristics.

The beverage bottle cooler can take a couple of forms. A very common approach to cooling bottles and containers is to use an upright

FIG. 12-6. Chest type bottle cooler. If self-service is desired, the sliding doors are available in glass or plastic. *Courtesy La Crosse Cooler Co.*

reach-in box with glass doors, mainly because the contents are quite visible. Such a box very easily lends itself to self-service. An alternate is the chest type bottle cooler, Figure 12-6. Its glass doors are horizontal and do not have the visibility and eye appeal of the vertical unit. It does have the advantage, however, of saving energy. The vertical box tends to lose a large amount of cold air every time it is opened. The heavy cool air spills out of the box quite readily. The horizontal case, however, will lose little cold air since the doors are in the top of the box. The trend seems to be to vertical boxes for self-service and the chest type for counter service. Both capillary tube and TXV systems are widely used in these boxes, in conjunction with blower type evaporator coils.

BEVERAGE DISPENSERS

Bottle coolers and reach-in display boxes are quite commonly used for the storage and sale of small containers of milk, soda, etc. Many food service operations buy beverages in bulk rather than prepackaged. Coolers for the storage and dispensing of milk by the glass, and soda systems, fall into this category.

Bulk milk dispensers, Figure 12-7, are refrigerators designed to hold

FIG. 12-7. Rear view of this bulk milk cooler shows the plate type evaporator and the louvers covering the condensing unit. Front view shows the temperature control knob and two dispensing faucets. This cooler holds two five gallon cartons of milk.

large containers, 5 gallons or more in size, of milk at a good serving temperature, usually around 40°F. Both plate type and blower coil type evaporators have been used for this application. Since the refrigeration load is relatively constant, the capillary tube finds wide application in this type of unit.

Soda systems fall into two general categories; post-mix and pre-mix.

PRE-MIX BEVERAGE SYSTEMS

In pre-mix systems, the mixing of a syrup and water, either carbonated or non-carbonated, occurs prior to dispensing the drink. In some systems, the mixing takes place at a central plant and 5 or 10 gallons containers are delivered to the food service line. In other systems, syrup only is delivered and mixing is done at the food service establishment, prior to being dispensed. In each case, the mixing is done at ambient temperatures and cooling is accomplished just prior to dispensing.

Figure 12-8 shows a counter top beverage dispenser for non-carbonated beverages. The pre-mixed beverage is poured into the reservoir at the top and the refrigeration system, consisting of the condensing unit

FIG. 12-8. Countertop beverage dispenser can be used for self-service or attendant service. *Courtesy Jet Spray Corp.*

below the reservoir and a bare tube evaporator in the reservoir itself, cools the beverage to serving temperature, usually around 40°F.

Figure 12-9 shows a system in which syrup tanks are delivered to the food server. An automatic filling system combines the proper amount of water and syrup and delivers it to the counter top dispenser for cooling to serving temperature.

The refrigeration systems used here are typically R-12 systems employing a capillary tube flow control. The automatic expansion valve also finds good use in this application.

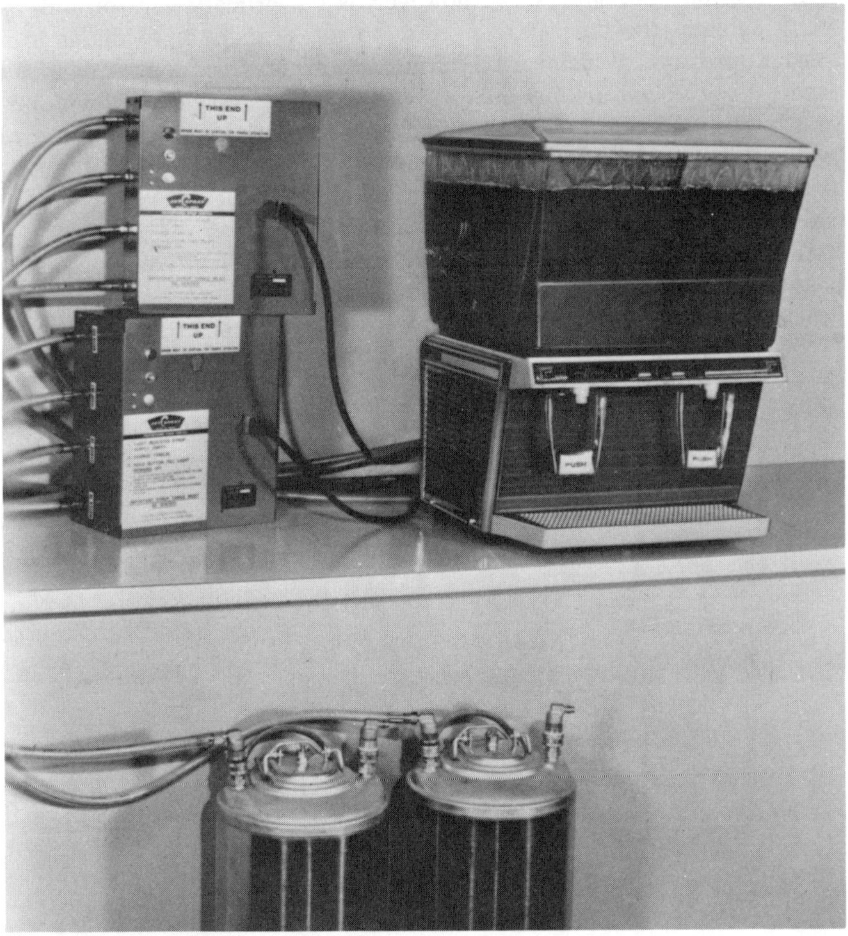

FIG. 12-9. Automatic filling system combines water and syrup and delivers it to a counter top dispenser. Note syrup tanks at bottom. *Courtesy Jet Spray Corp.*

AUTOMATIC EXPANSION VALVE

The automatic expansion valve, Figure 12-10, is infrequently employed in commercial refrigeration. It was developed fairly early in the history of refrigeration, and enjoyed some early popularity, but has now been largely replaced by the TXV and cap tube.

The principle of operation of the automatic expansion valve or AXV is very simple. It is designed to maintain a constant pressure in the evaporator. This is similar to the evaporator pressure regulator used when multiplexing. The AXV, however, is the primary control and, as such, regulates the flow of refrigerant to the evaporator to maintain a constant pressure. Whenever the term *constant pressure* is used, you can equate it to *constant temperature*. The AXV, by its very design, will prevent beverages from freezing in the evaporator, even if a loss of control by the thermostat or a refrigerant leak should occur. This non-freezing characteristic is one of the things that recommends the AXV in beverage cooling applications.

In discussing both the capillary tube and the TXV, the use of superheat was mentioned. The AXV totally ignores the significance of superheat and is adjusted strictly on the basis of evaporator pressure. In fact,

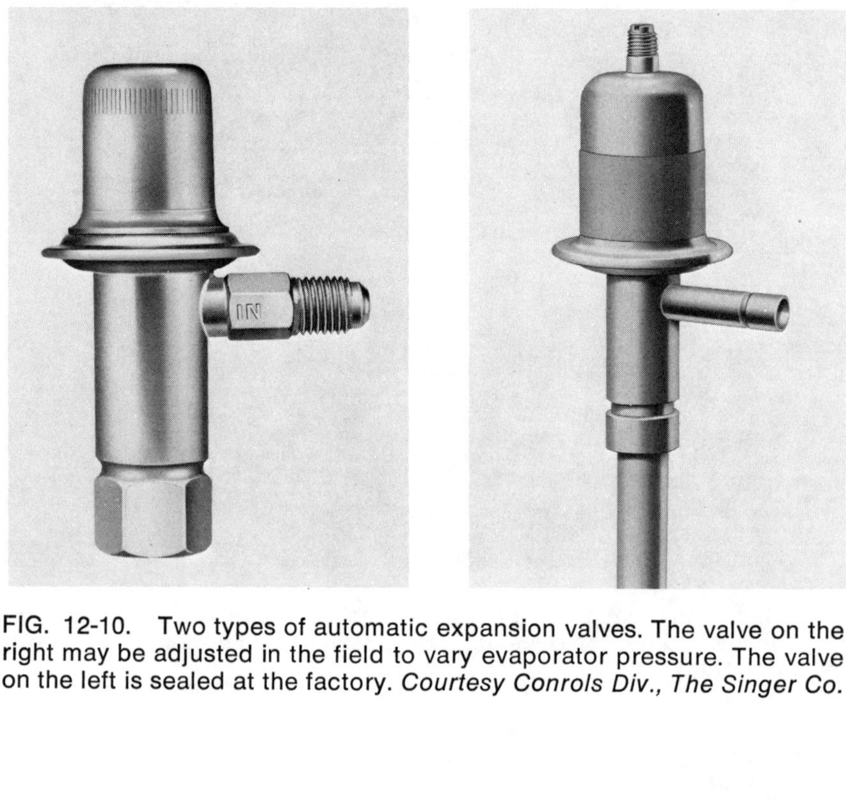

FIG. 12-10. Two types of automatic expansion valves. The valve on the right may be adjusted in the field to vary evaporator pressure. The valve on the left is sealed at the factory. *Courtesy Conrols Div., The Singer Co.*

the major shortcoming of this valve is its action under varying load, the area where the TXV is most outstanding. As the load on the evaporator increases, the evaporator pressure will tend to increase, since the refrigerant will boil off at a greater rate. The TXV will open, admitting more refrigerant to take care of the increasing load. The AXV, however, does just the reverse and closes down, restricting refrigerant flow until the evaporator pressure drops down to the valve's set point. This set point, on beverage cooling systems, is usually around 30 psig. Of course, as low side pressure drops off, the AXV will open and admit more refrigerant to boost the pressure back up again. You can see why the valve isn't quite as popular as it once was. Under low load conditions, the valve would tend to open wide, possibly permitting liquid refrigerant to flow all the way back to the compressor, causing all sorts of problems. In a beverage cooling application, where the load doesn't vary greatly, the AXV is fine. In those applications where widely ranging loads are encountered, it has been supplanted by the TXV.

POST-MIX SYSTEMS

In pre-mix systems, the beverage is mixed prior to dispensing. In post-mix systems, Figure 12-11, only syrup is provided by the manufacturer.

FIG. 12-11. Refrigerated post-mix dispensing unit has solenoid type faucets for blending the proper ratio of carbonated water and syrup.

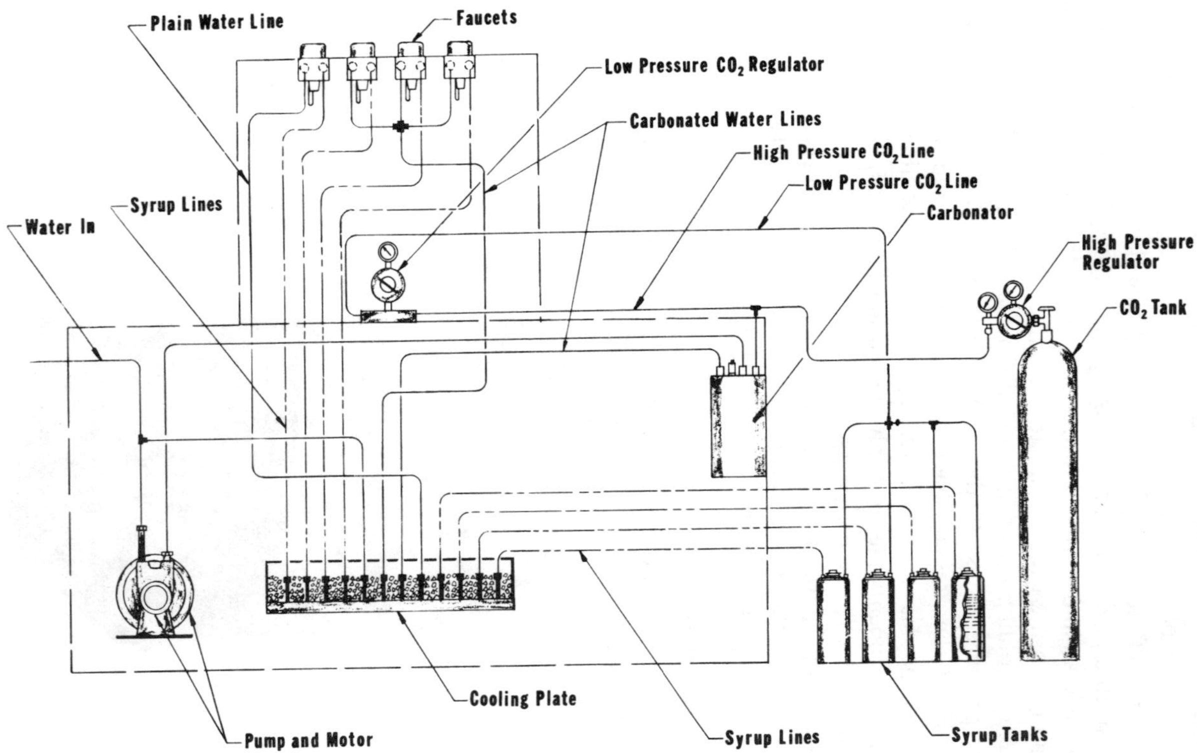

FIG. 12-12. Schematic of a post-mix system that uses a cooling plate set in a bin of flaked ice. Faucets can be either solenoid or mechanically operated. *Courtesy Sunbeam Mile High Equipment Co.*

Carbonated water and the syrup are delivered to a dispensing unit and the actual mixing of the drink is accomplished in the cup at the point of purchase. The equipment necessary to provide a finished drink, in this type of system, is more expensive than with the pre-mix arrangement, but the profit on each drink is greater.

The essential components of the post-mix system are the syrup tanks, the carbonator, and the dispensing unit, Figure 12-12.

The carbonator has two major components: the tank, Figure 12-13, and the pump. The low pressure water is connected to the intake of the pump and discharged into the tank which is under 100 psig pressure from the remotely located carbon dioxide bottle. Since the city water pressure is far less than that of the carbon dioxide, a pump is needed to force the water into the tank. The high pressure carbon dioxide, mixing with the water, produces carbonated water or soda.

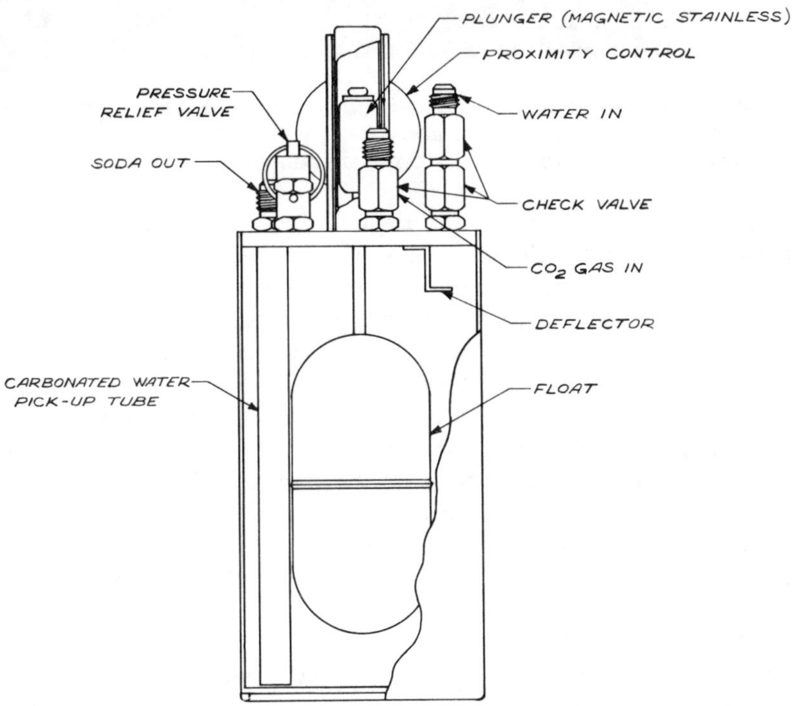

FIG. 12-13 Cutaway of the carbonator shown in Figure 12-12. High pressure carbon dioxide gas mixes with water delivered by the pump at pressures exceeding 100 psig. The carbonated water is then delivered to the dispensing unit where it is blended with syrup. *Courtesy Sunbeam Mile High Equipment Co.*

FIG. 12-14. Back room scene shows carbon dioxide tank with its high pressure regulator, syrup tanks, and low pressure regulators. Low pressure carbon dioxide forces syrup to the dispenser through plastic tubes.

Pressure of the carbon dioxide, Figure 12-14, forces both syrup and carbonated water through parallel lines to the dispenser where the proportion of syrup to water is regulated by electrically operated solenoid valves or mechanical valves with specifically selected orifices. Sweetness of the drink can be regulated by adjusting the orifice size.

Prior to entering the dispensing unit, both the syrup and the soda are usually not refrigerated. The dispensing unit has, along with the syrup metering apparatus, a refrigeration system. In some cases, the evaporator described for pre-mix soda systems will be used or perhaps, alternatively, a cast aluminum block with direct expansion cooling might be used. In either case, the condensing unit is quite similar to those previously described. Both remote and self-contained equipment can be encountered.

FROZEN BEVERAGE MACHINES

The term *frozen beverage* is used to collect a number of similar refrigerating machines into one category. The products of these machines are *frozen custard, soft ice cream, thick shakes,* or *heavy frosteds,* which are dairy products, and *slush ice* or *Italian ice,* which are non-dairy. The differences between the products depend upon the mixture put into the freezing machine. The equipment itself, however, is quite similar, regardless of the product, Figure 12-15.

FIG. 12-15. Dispenser typical of those used to produce soft frozen products such as shakes, ices, and carbonated slush in the range of 20 to 22°F. *Courtesy Sweden Freezer Mfg. Co.*

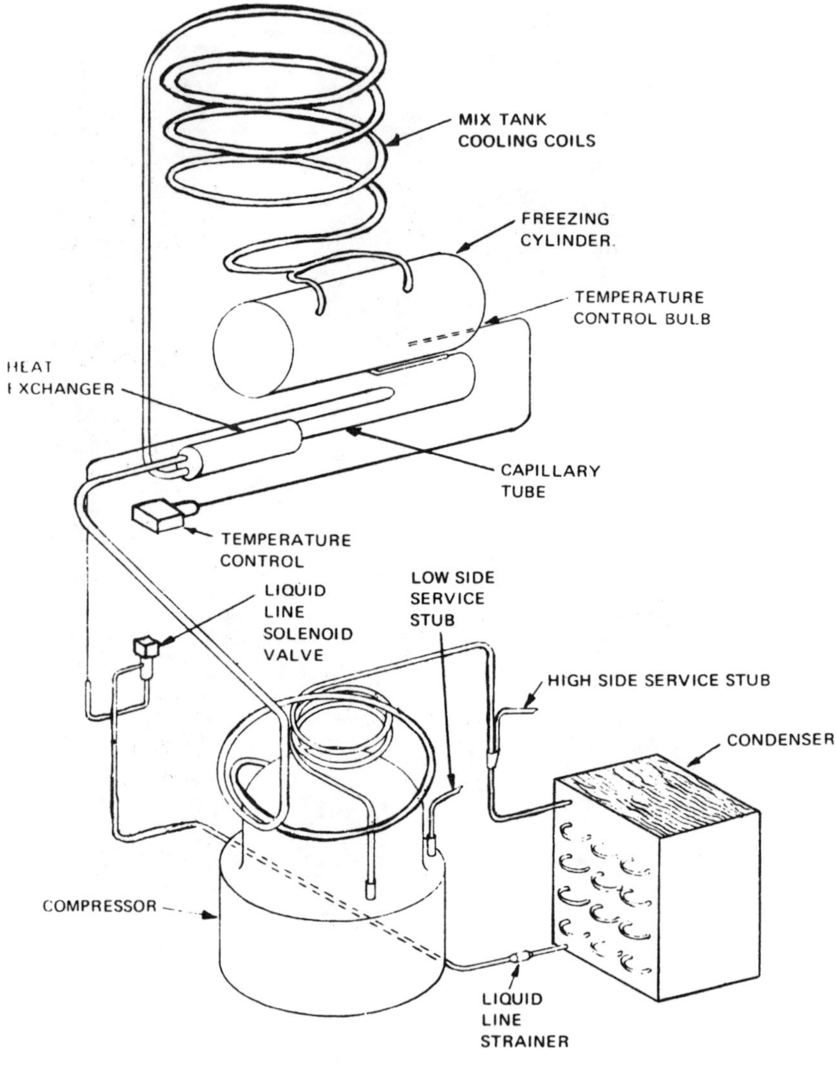

FIG. 12-16. Refrigeration system for a frozen beverage machine. Note that there are two evaporators: one precools the mixing tank while the other freezes the mix prior to dispensing. *Courtesy Sweden Freezer Mfg. Co.*

The condensing units are the same, in all respects, as those encountered in earlier applications. The evaporator is, again, the only unique part of this equipment. Usually, these machines have two evaporators connected to a single condensing unit, either remote or self-contained. The first evaporator is used to pre-cool the mix before it enters the second evaporator known as the freezing chamber, Figure 12-16. The product is eventually cooled to a temperature of about 20 to 22°F for final dispensing. Hard ice cream, in comparison, is typically served at a temperature of about 5 to 10°F. It is quite common to have the pre-cooler controlled by a capillary tube and the freezing chamber controlled by an AXV. Remember, the AXV is designed to maintain a constant pressure, hence temperature, in the evaporator, thereby preventing the product from freezing solid in the machine.

A fairly recent innovation is a machine designed to dispense carbonated, frozen *slush,* Figure 12-17. This machine is actually a combina-

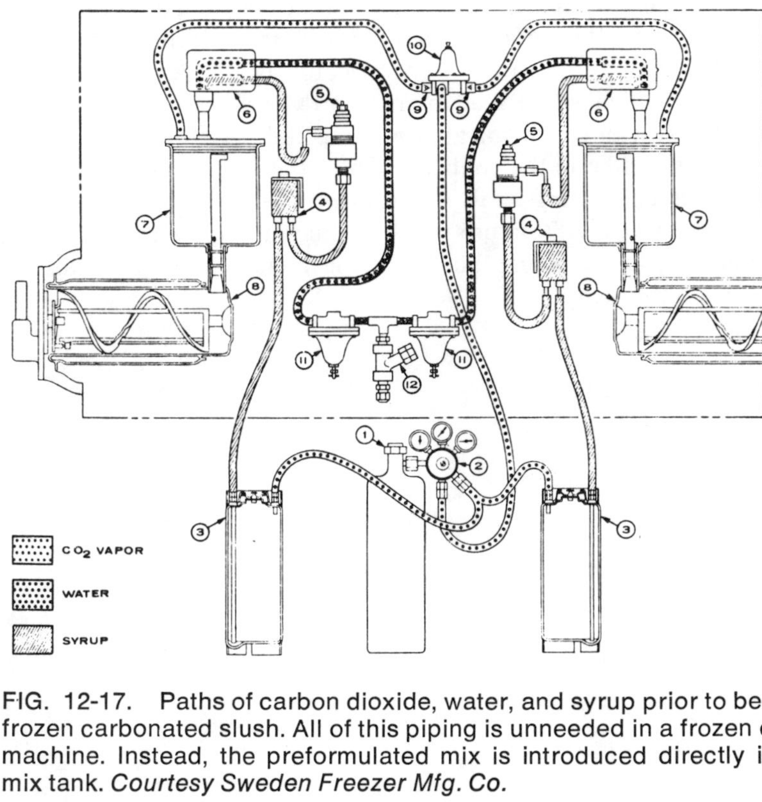

FIG. 12-17. Paths of carbon dioxide, water, and syrup prior to becoming frozen carbonated slush. All of this piping is unneeded in a frozen custard machine. Instead, the preformulated mix is introduced directly into the mix tank. *Courtesy Sweden Freezer Mfg. Co.*

tion soda system, since the product is carbonated, and *shake* machine, since the product must be frozen to a slushy consistency.

While the refrigeration system should pose no major problem to the service technician, these machines are somewhat unique. The typical indication of a malfunctioning machine is not an increase in temperature, but rather a deterioration in the appearance of the product. The change in appearance may be related to a refrigeration problem but, interestingly enough, this is true in less than half the service calls on these machines. More commonly, the equipment is being overworked, that is, being called upon to deliver more servings of a product than it was designed to deliver. At other times, the product mix has varied from batch to batch or possibly, after cleaning, the machine was improperly reassembled. None of these problems are related to the refrigeration system.

It is a good idea for the serviceman, not familiar with these machines, to telephone and consult with the manufacturer's representative before he gets into too much trouble.

Finally, a vital part of any serviceman's tool kit is a library of technical manuals describing equipment that is somewhat unique and not encountered everyday, yet often enough to warrant having the technical information available. Falling into this category would be shake machines, ice makers, about which you'll read more shortly, soda systems, and any other equipment with which you are not completely comfortable. Usually, the uniqueness will be in the configuration of the evaporator, but it may also be in pressure settings, valve adjustments, water levels, flow rates, etc. In many cases, the literature you have may not apply specifically to the equipment you are facing, but it may have enough general information to see you through.

13

Ice Makers

While the day of the iceman may be gone and long forgotten, use of ice is not. In fact, ice is as popular as ever. The use of ice cubes for cooling drinks and flake ice for food preservation during shipment and display is so widespread that the commercial serviceman who cannot perform at least the simplest service on ice making equipment is at a great disadvantage.

The main difference between the ice industry of a generation ago and that of today is in the size and location of the equipment. Today, small ice makers, with capacities as low as 100 pounds, and large systems, making several thousand pounds every 24 hours, are located on the premises of the user. In years past, central ice plants would make blocks and distribute either blocks, cubes or crushed ice as customers requested.

FIG. 13-1. Modern ice makers have closed storage bins to prevent contamination. *Courtesy Queen Prod. Div., King-Seeley Thermos Co.*

While the central ice plant is not entirely defunct, it has diminished in importance over the years.

In this chapter we'll look at cubers and flakers to see how they operate and develop a systematic inspection procedure that will, in most cases, lead to the solution of service problems.

One of the main themes of this book has been the similarity of refrigeration equipment used in various applications. In general, the design of the refrigeration and electrical circuitry was independent of the manufacturer and was basically related to the application of the equipment. In looking at ice making equipment, Figure 13-1, this thesis breaks down somewhat. Each of the thirty or more manufacturers of cubers and flakers has unique design features. It is this variety in the making of ice that, quite frankly, makes ice machine service a challenge and poses the greatest problems to the man in the field. The problems are certainly not insurmountable, but they do remove ice machine service calls from the realm of the routine.

FIG. 13-2. This cube maker, typically found in a small restaurant or tavern, is rated at 200 lb of ice in 24 hours. *Courtesy Queen Prod. Div., King-Seeley Thermos Co.*

ICE MACHINES

The output of ice machines is rated in pounds of ice per 24 hours, Figure 13-2.

The most widely used machines will produce either cubes or flakes. Cubed ice is available in many different shapes, dependent upon the manufacturer of the machine and the process he uses. A fun game is to attempt to identify the manufacturer of the ice you find in your drink at a local tavern or restaurant. After a few years of ice machine service, you'll probably know a good number of them. The trend in recent years has been away from the large, bulky cube and to a smaller cube, more of which can fit in a glass. The majority of machines sold today make small cubes.

One of the main selling points of commercial ice makers is the purity and clarity of the cube. These qualities are achieved by continually pumping the water over an evaporator during the freezing cycle. This continuous circulation prevents air from being entrained in the ice, unlike home refrigerators where the water is stationary and air bubbles cause cloudy cubes, and also prevents impurities from being trapped within the ice. An interesting experiment is to add food coloring to the water in an ice machine. After a batch of cubes has been made, examine them for color. You'll find that the coloring doesn't stay in the cube. This is typical of most impurities. An analysis of the water from a melted cube will show a high degree of purity even though the water supply to the machine has a high mineral content. Again, this can be compared to the cubes from a home freezer where color and unpleasant taste can be detected in the ice.

While cubers come in many varieties, the flakers, fortunately, are more limited in design. The basic approach is to have a drum submerged in a reservoir of water. The drum acts as an evaporator and ice will form on its surface. The film of ice is scraped off and compressed or is cracked and wiped off, depending on the design. We'll take a closer look at flakers shortly. Since the water in the reservoir is generally not moving, most flaked ice is cloudy and bears traces of discoloration, odor, and taste associated with the water supply of the locale. If flaked ice is to be used in drinks, rather than in food preservation, it is not uncommon to have some form of water treatment apparatus in the supply line to insure palatable ice in areas of poor water quality.

OPERATION OF ICE CUBERS

A good way of understanding how ice cubes are made is to envision the ice machine as having three subsystems combined into one larger system. The subsystems are: water, refrigerant, and electrical.

Water Subsystem

All ice cubers circulate water over the face of an evaporator, Figure 13-3. The temperature of the evaporator is below freezing, therefore, some of the water will turn to ice and remain on the evaporator surface. The rest will fall into a sump where it is drawn into a pump and recirculated back to the evaporator face once again. The water pump in constant operation, while ice is being formed.

In addition to a water pump, the water subsystem must have tubing, usually plastic, to carry water to the evaporator, Figure 13-4. This tubing must be clean and unobstructed, if the water is to flow properly. Also required is a means of distributing the water evenly over the entire face of the evaporator. A stationary spreader bar, with small diameter holes, or an oscilating bar, also with holes, will spread the water evenly, assuring uniform development of the cubes.

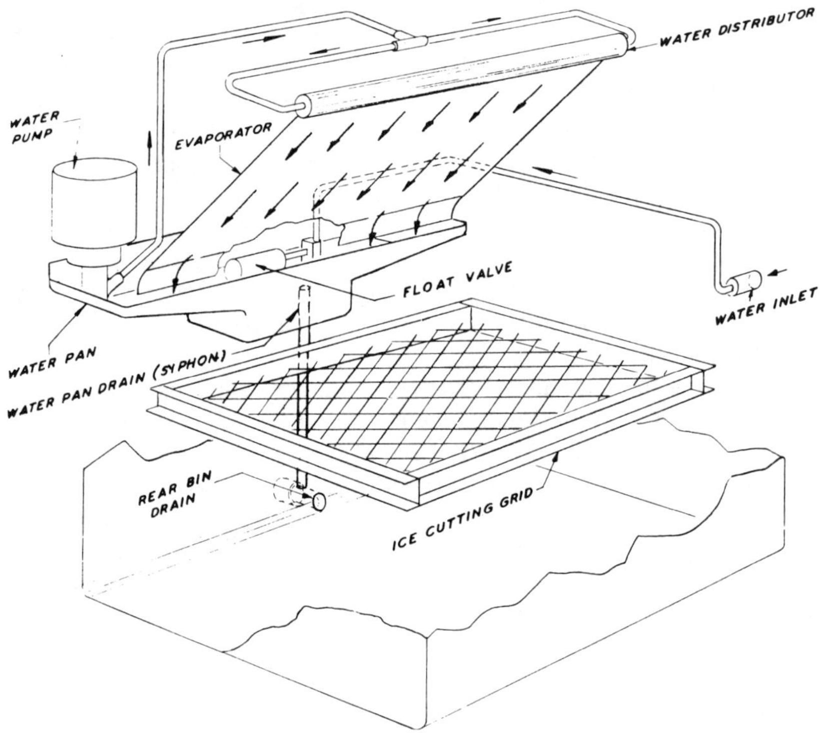

FIG. 13-3. Typical water circuit for an ice maker that forms a slab on a plate type evaporator. The water is continuously recirculated by the pump and, as the water freezes, more is admitted to the water pan by the float valve. At the end of the cycle, the mineral-laden excess water is removed by the syphon. *Courtesy Frigidaire Div., GM Corp.*

The cubes are generally made by a batch process, that is, a fixed number of cubes are made during every cycle. A predetermined amount of water must be present at the beginning of the freezing cycle. In some machines, the total amount of water needed for a batch of cubes is contained in the sump at the beginning of the cycle. In other machines, water is metered into the sump, by a float valve, as ice forms on the evaporator. In either case, sufficient water must be present to permit full size cubes to develop.

A portion of the water will be converted into cubes and some will remain behind. The remaining water plays an important role in assuring troublefree operation. As previously described, minerals and other impurities in the water do not freeze into the ice, but rather stay behind with the unused portion of the water. At the end of every freeze cycle, the remaining water must drain off, taking the accumulated impurities with it. If this does not happen, the mineral content will build up and clog the water passages, causing early breakdown of the equipment. All

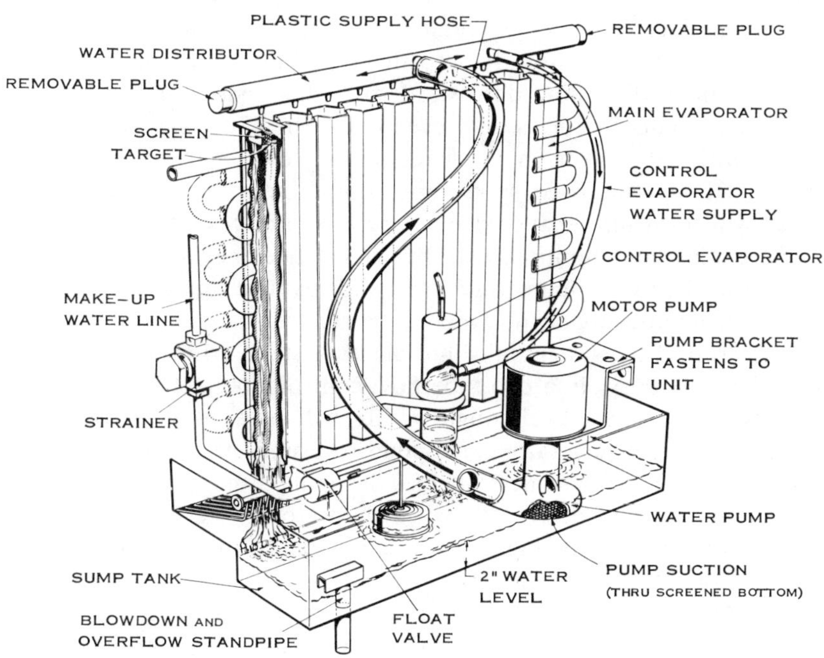

FIG. 13-4. Another style cuber with the same elements found in Figure 13-3: pump, water distributor, sump tank and float valve. The tubing carrying the water must be clean and unobstructed. *Courtesy York Div. Borg-Warner Corp.*

machines have a mechanism for draining all or most of the remaining water at the end of each cube batch.

During the freeze cycle, that is, while ice is being formed, the water system is in operation and water is being pumped, through tubing, and distributed across the face of the evaporator. During the harvest cycle, that is, while the ice is being removed from the evaporator, the water subsystem is idle.

Refrigeration Subsystem

Fortunately, as we look at the refrigeration system, we are once again on familiar ground since it is similar, in most respects, to those we have seen before. The major difference is the shape of the evaporator. The evaporator, being the cold surface in the refrigeration system, is the surface upon which the water will freeze and its shape will determine the shape of the cube.

Some ice machines have a flat plate evaporator, Figure 13-5. Water flows over the inclined plate and a slab of ice, of a prescribed thickness, is formed. Other ice machines have a series of cubical or cylindrical compartments molded into the evaporator, in either large, medium or small sizes. Water is directed upward into these cavities, if the plate is horizontal, or allowed to flow downward over the cavities if the plate is vertical. Ice gradually forms and, eventually, the cavities become filled.

Since other ice makers have water flowing over vertical evaporator plates that have embossed spots or sections, these spots become cold and ice will form into shapes defined by the pattern of the cold spots. There are other configurations used but, in essence, they are all the same, namely a cold surface upon which ice builds as water plays over it.

Once the uniqueness of the evaporator has been examined and understood, the rest of the system falls into place. A standard condensing unit, with a TXV or possibly a cap tube flow control, completes the picture. Typically, R-12 is used, although R-502 will be seen upon occasion. Low side pressures generally correspond to evaporator temperatures of about 15 to 20°F, with some machines dropping a bit lower. Usually, at the beginning of the freeze cycle, when the water is relatively warm, the low side pressures will be higher. As the ice builds, the evaporator pressure will drop into the design range.

In the absence of any other information, it is a good idea to adjust the refrigerant flow control and/or the refrigerant charge to maintain a superheat of about 10°F. Even though you might not know what the operating pressures should be, adjustment of the system to this superheat will usually result in proper operation. Incidentally, this adjustment should only be made after ice has begun to form on the evaporator surface, that is, about 10 or 15 minutes into the freezing cycle.

The best way to determine the proper pressures is to consult the manufacturers' operating and maintenance instructions. Much detailed information is available from manufacturers and a large, 4-inch ring binder, containing instructions for the machines you commonly service, is a useful addition to the service truck.

During the harvest cycle, the ice is freed from the evaporator and allowed to fall into the storage bin or, in the case of flat plate units, the slab of ice is released to fall onto a wire grid, Figure 13-3, where it is cut into cubes.

The most common means of harvesting is the hot gas method. During

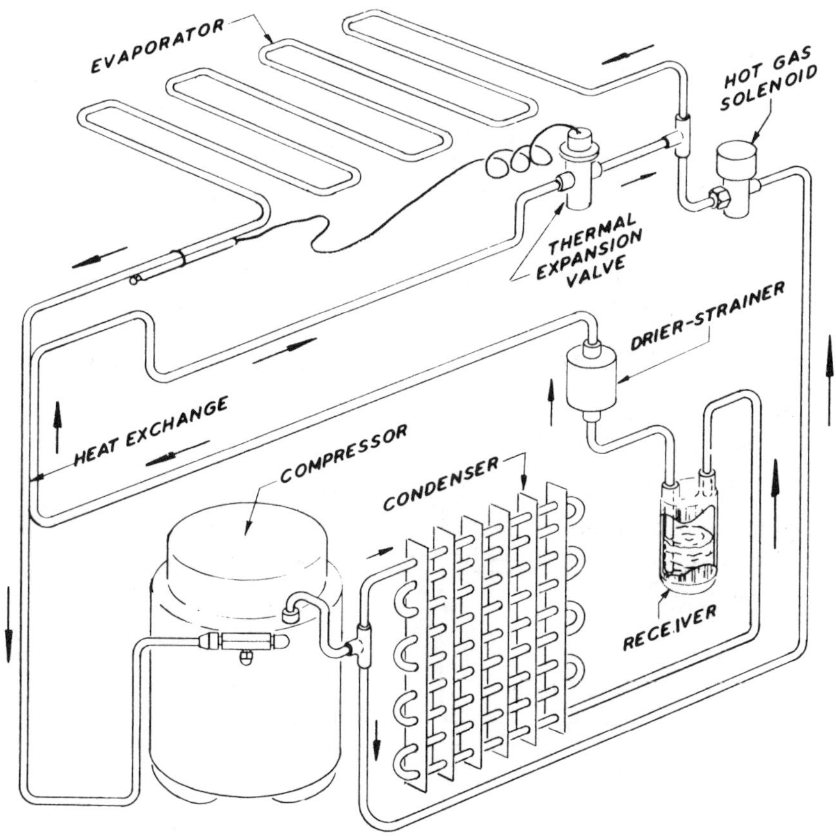

FIG. 13-5. Refrigeration system for an ice cube maker employing a plate type evaporator. The plate is soldered to the evaporator tubing. *Courtesy Ross-Temp, Div. of Schneider Metal Mfg. Co.*

the harvest cycle, the compressor continues to run. However, the hot gas from the compressor is diverted to the evaporator by means of a solenoid valve. This, of course, is quite similar to the hot gas defrost system used with freezers. After the hot gas warms the evaporator, the weight of the cubes will pull the ice free.

Harvesting is a rather short cycle, generally on the order of 5 to 10 minutes. The freezing cycle, by comparison, can last as long as an hour

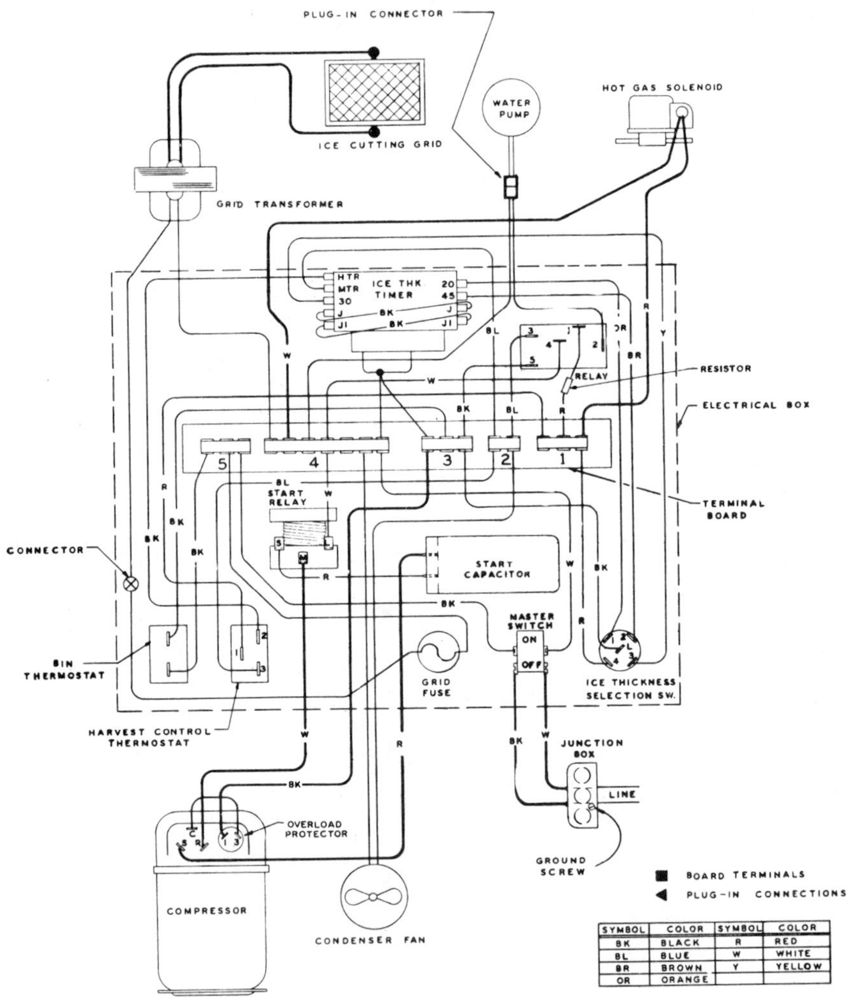

FIG. 13-6. This manufacturer's wiring diagram combines a pictorial diagram , to help locate components, and a schematic diagram, to help in electrical troubleshooting. These diagrams, and the manufacturer's

with smaller machines and about 20 to 30 minutes with some larger machines. After all the ice has fallen away from the evaporator, the harvest cycle terminates and the freeze cycle begins once again.

Some harvest systems use electric heaters to warm the evaporator. This method is not nearly as popular as the hot gas system. Another manufacturer uses a warm water bath, directed to the reverse side of the evaporator, to loosen the cubes. These alternate techniques are not widely used but, when encountered, can cause confusion unless you know of their existence.

Electrical Subsystem

There is little doubt that the most preplexing problems arise in the electrical control subsystem. It is in this area that manufacturers' wiring

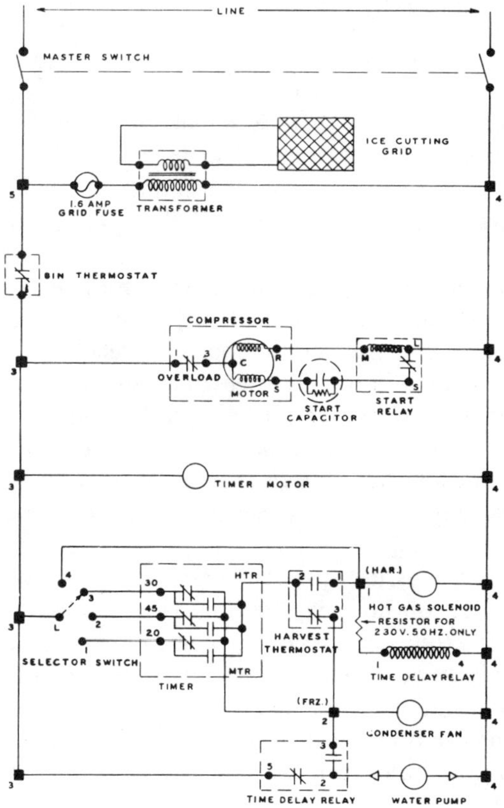

description of the sequence of operation, establish how a machine should operate and point to the source of a malfunction. *Courtesy Ross-Temp, Div. of Schneider Metal Mfg. Co.*

diagrams, Figure 13-6, and an ability to read and interpret them are mandatory. The controls start, stop, and regulate the ice making process and must be understood if the serviceman is to be a complete technician.

Smaller ice makers usually run on 115 volts while larger sizes, generally producing over 400 lbs. per day, run on 230 volts. Occasionally, an ice machine will have a combination of 230 volt and 115 volt components, which can be tricky if you don't look for it.

Line power to most machines is controlled by the bin thermostat. This stat senses the temperature of the bin and shuts down the unit when the temperature drops due to the bin being full. In some designs, a mechanical device trips when the bin has been filled to the desired level. Most commonly, however, a thermostat is used. If the ice machine is located in an area of low temperature, say below 50°F, the bin stat may give a false indication, shutting the machine down before it is full or failing to start the machine after ice has been removed. Good practice dictates that ice makers be located in heated areas having temperatures above 55°F year-round.

Once past the bin control, there is a maze of wires and controls that is more complex in some machines than in others. Typically, they are grouped, by function, as the controls that initiate the harvest cycle, controls that terminate the harvest cycle, and the controls that monitor the freeze cycle.

At the end of the freeze cycle, some device must recognize that a full set of cubes is ready to be harvested. Many machines use a control that senses the suction pressure which, of course, is related to the load on the evaporator. The load is low when ice thickness is great, therefore, it would seem feasible to relate ice thickness to low side pressure. It is not uncommon to have this pressure control energize a time clock that allows the freeze cycle to continue for several minutes, then initiates the defrost cycle. This clock may also have a setting to control the duration of the harvest cycle.

Since evaporator pressure and temperature are related, a thermostat mounted on the suction line may be used to sense the depth of the ice build-up. Flat plate units may have a thermostat located above the ice slab. As the slab becomes thicker, it eventually comes in contact with the suspended thermostat, initiating the harvest cycle.

Several designs use a motor driven sensing arm that rotates above the ice. Upon hitting the built-up ice, the arm trips a miniature switch that terminates the cycle.

Another machine has a fixed amount of water in a reservoir at the beginning of the freeze cycle. When the water level falls, as a result of the water forming ice on the evaporator, a switch is tripped initiating the harvest cycle. As we get deeper into the subject, you can see that there are many ways of skinning a cat and it seems that all of them have been tried in the ice machine field.

When all the cubes have fallen from the evaporator, the harvest cycle must be terminated before a new freeze cycle begins. As mentioned before, a clock timer can terminate the cycle. Thermostats, sensing evaporator temperature, are also used. Sometimes the stat that initiated the harvest cycle also terminates it, but we will often see an entirely different stat doing the job. In some machines, a mechanical switch trips when the ice has left the evaporator. This system will work with slabs of ice or with cubes held together by a thin web prior to their breaking apart as they fall into the bin. Whatever the method, the termination device must start the refrigerant flowing in the right direction and restart the water pump for another freeze cycle.

TROUBLESHOOTING CUBERS

The multiplicity of ice cube maker designs, and their many variations, make troubleshooting a challenging task. Since, to be effective, you must understand the principles of operation of each of these designs, a complete library of manufacturers' installation and operating instructions should be kept near at hand. Lacking technical literature, the average serviceman can effectively solve many ice machine problems, if he has the confidence to delve in and try.

Keep in mind that there are three major subassemblies that make up the typical cuber: water, refrigeration, and electrical. Problems occurring in the first two categories are relatively easy to pinpoint and to repair. In most cases, a visual inspection will reveal the problem with the water circuit. An analysis, similar to that performed on reach-in and walk-in refrigerators, will usually detect most of the problems with the refrigeration system.

Recognize that there are two modes of operation of the ice cuber: the freezing cycle, during which water is sprayed over the face of the evaporator and freezes into ice; and the harvest cycle, during which the evaporator is warmed, permitting the formed ice to separate from the evaporator and fall into the bin or onto cutting wires. By continually asking, "What *is* actually happening?" and "What *should* be happening?" you can usually isolate the problem fairly quickly. Of course, you must know what should be happening at a particular point in the operating cycle. This will take clear, logical thinking and confidence that you can figure out the problem. Failure to solve a problem is due more often than not, to being convinced, from the outset, that the solution is beyond you. This is not true. Hang in there and, most importantly, THINK.

Common Problems	Common Causes	Uncommon Causes
Unit completely inoperative	Blown Fuse Defective Bin Stat Off on Bin Stat	
Low ice making capacity	Dirty Condenser Dirty Evaporator	Defective compressor Water leaking into sump during cycle Leak in hot gas valve Water leaking from unit Defective expansion valve
Thin cubes	Improperly set or defective thickness control	
Thick cubes or failure to go into harvest cycle	Improperly set or defective thickness control	
Short cycling compressor	Dirty condenser Low refrigerant charge Defective relay or capacitor	

This table is quite sketchy, however, it does point out several major problems common to most cubers and their causes. With some few exceptions, you will be readily able to handle any problems coming your way.

PREVENTIVE MAINTENANCE

Because of their added complexity, ice makers tend to require more service than other refrigeration units. One way of reducing repairs is to diligently follow a schedule of preventive maintenance.

Preventive maintenance can easily be overdone, particularly if the schedule requires disassembly of components or the moving of parts or wires that may be perfectly all right when you start, but may become damaged during inspection.

Apart from visual inspection of readily-seen components, the only preventive maintenance that is really needed is to lubricate the condenser fan motor and water pump motor, if possible; to clean the surface of an air-cooled condenser or the water passages of a water-cooled condenser; and to run chemical cleaner through the water subassembly of the cuber itself. These procedures will go a long way toward increasing the useful life of the equipment.

Ice machine cleaner is manufactured by a number of companies and is designed to remove accumulated scale. This scale, if it becomes excessive, will coat the evaporator, decreasing its efficiency, and might prolong the time that the cubes cling to the evaporator surface during the harvest cycle. These cleaners are quite easy to use. In fact, many ice maker manufacturers encourage their use by equipping the machine with a Wash Cycle. During this Wash Cycle, only the pump operates, to circulate water and cleaner through the water passages. Annual or even semiannual cleaning is a worthwhile exercise.

ICE FLAKERS

Typical flakers, Figures 13-7, 13-8 and 13-9, consist of a refrigeration system, of the type we have seen before, a water feed mechanism, and a scraper device.

FIG. 13-7. Flaked ice machine with a self-contained bin. *Courtesy Queen Prod. Div., King-Seeley Thermos Co.*

FIG. 13-8. Two flaked ice machines, with different ice making capacities, mounted on a common bin. With this arrangement, bin capacity need not be limited to the flaker output. *Courtesy Queen Prod. Div., King-Seeley Thermos Co.*

FIG.13-9. Flaker, bin and dispenser are combined in a compact countertop unit. Other units dispense water, as well. *Courtesy Queen Prod. Div., King-Seeley Thermos Co.*

The only unique feature of the refrigeration system is the cylindrical shape of the evaporator. Refrigerant tubes are either bonded to the outside or the inside surface of the evaporator, depending upon the manufacturer's design, Figure 13-10. Water either fills the cylinder or surrounds it and is kept at a constant level by the water feed mechanism.

Flaked ice is produced on a continuous basis rather than the batch method favored with cubes. The ice forms, in thin layers, on the surface of the cylinder and is scraped off by a rotating worm or auger. The scraped ice is then forced out of a passageway in the form of flakes.

Flakers are easier to troubleshoot than cubers, mainly because they are less complex. The refrigeration cycle, with either a TXV or cap tube, usually uses R-12 refrigerant. Typical low side pressure is on the order

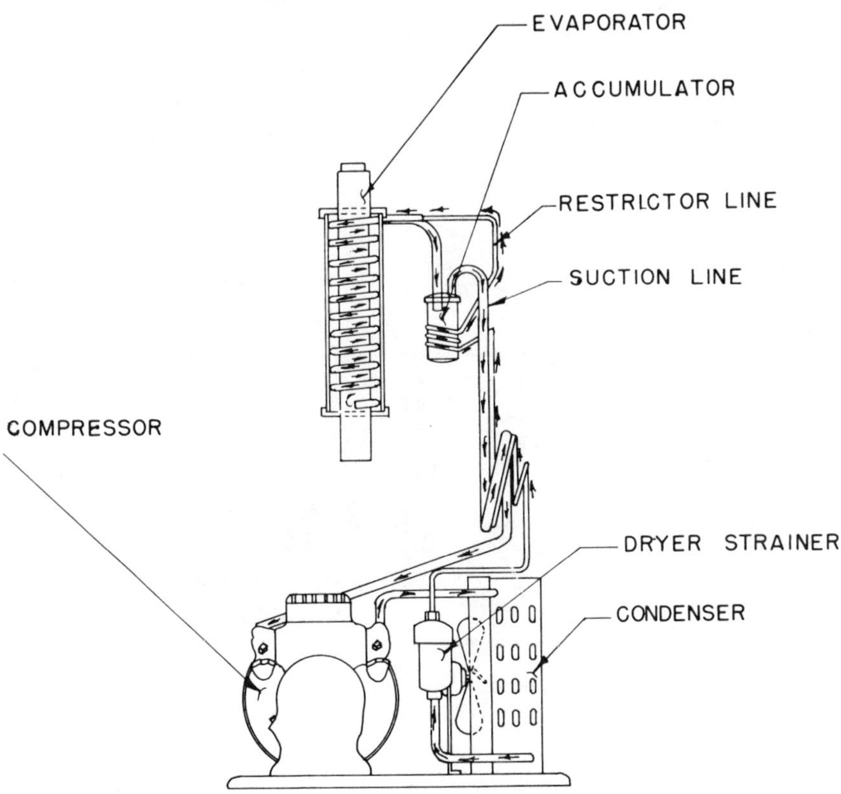

FIG. 13-10. The shape of the evaporator and the inclusion of an accumulator are characteristics of flaked ice machines, Note that this system employs a capillary tube (restrictor line). *Courtesy Queen Prod. Div., King-Seeley Thermos Co.*

of 11 to 15 psig, once the system has achieved stable operation. A liquid/suction heat exchanger or an accumulator might be found on some models but, in general, the refrigeration side of the system is quite straightforward.

The water feed mechanism, Figure 13-11, employs a float valve designed to maintain the water in the evaporator at a fixed level. The water must be maintained at the design level but, beyond that, nothing complex is involved.

The scraper mechanism, Figure 13-12, is another matter. On the face of it, a set of gears, rotating at only 60 revolutions per minute, would seem to present no major problems. The truth of the matter is that probably as many problems arise in the auger assembly as in the rest of the machine combined. It is not uncommon to have the auger bearings wear ever so slightly, just enough to cause the auger to jam. When this happens, the bearings must be replaced.

Bearing wear is not the easiest problem to diagnose since it is almost imperceptible in many cases. After a while, the problem is pinpointed by either the process of elimination, there being no other apparent cause of the jamming, or by developing a fine sense of touch. A gage tester would

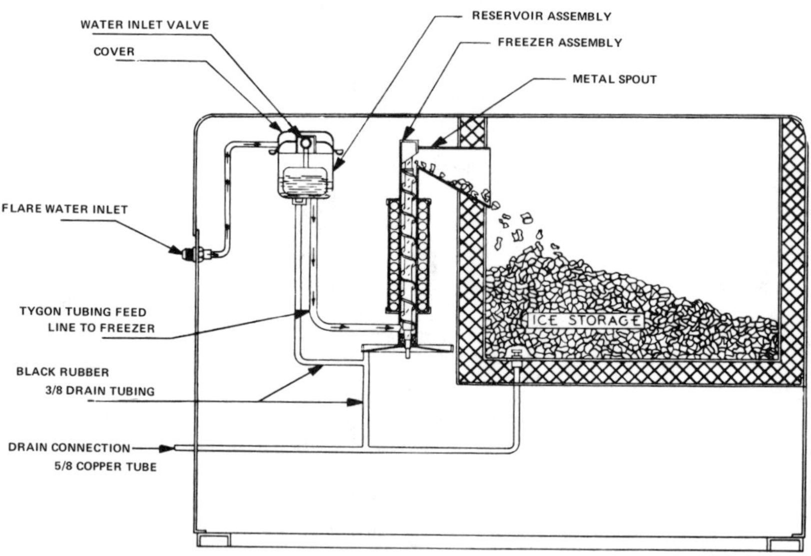

FIG. 13-11. While simpler than that of a cuber, the water delivery system must be above the top of the auger to assure rated ice making capacity and smooth auger operation. *Courtesy Queen Prod. Div. King-Seeley Thermos Co.*

be helpful, but few servicemen have such an instrument. If they do, it is seldom accurate enough to read the slight amount of wear that will cause an auger to bind.

FLAKER OPERATION

As with a cuber, a bin thermostat is usually used to initiate the operating cycle of the flaker. During the start-up period, the compressor will be energized and the suction pressure will gradually fall as the water in the evaporator gets chilled and begins to form a layer of ice on the surface of the evaporator. The auger will be in motion from the start.

When the ice on the evaporator is sufficiently thick, it will meet the edge of the auger and be scraped free. The auger will collect the flakes and transport them to the top of the evaporator and into the ice chute. The accumulation of ice will gradually force previously scraped flakes through the chute and into the bin.

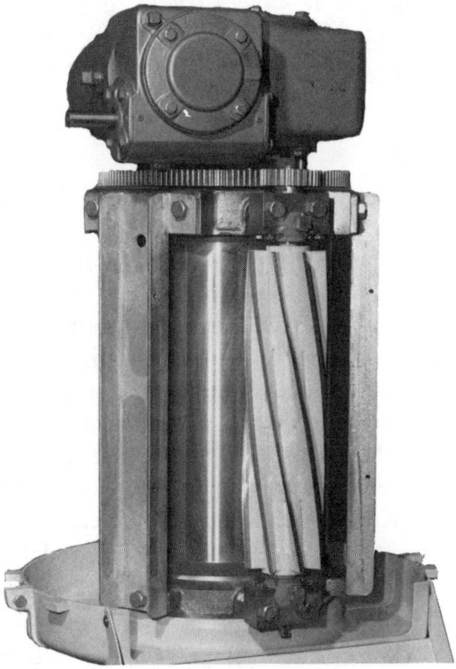

FIG. 13-12. Here is another popular method of harvesting flaked ice that is used in machines producing up to 12,000 lb per 24 hours. Water is sprayed on the stationary evaporator drum where it quickly freezes. The ice is chipped of the drum by a rotating cutter that circules the periphery. *Courtesy Flakice Corp.*

Evaporator pressure, which is dependent upon refrigerant charge and high side pressure, is a critical factor in flaker operation. If the pressure gets too low, the ice will become too hard and offer excessive resistance to the auger. This may overload the auger motor, causing it to cut out on safety, or may increase the wear on the auger bearing and lead to premature failure. Evaporator pressure, as previously mentioned, should be on the order of 11 to 15 psig. Substantially lower pressure will assuredly create problems.

If the water is not up to the design level, two things will result. First, the output of the machine will decrease because only a portion of the evaporator is producing ice. Second, with less water in contact with the evaporator, a lower refrigeration load will exist. This, in turn, means that the suction pressure will tend to drop. As we've seen, low suction pressure means a cold evaporator and tends to overload the auger as it tries to scrape the harder than normal ice from the evaporator surface.

Troubleshooting the system boils down to analyzing three subsystems: refrigeration, water supply, and auger. Again, a thoughtful approach, asking the question, "What is happening?" and "What should be happening?" will carry the day. There is nothing here that the good mechanic can't handle.

PREVENTIVE MAINTENANCE

The preventive maintenance program prescribed for cubers also applies to flakers. Motor lubrication, cleaning the condenser, and cleaning the water passages pretty much sums it up. The flaker does have one peculiarity the cuber does not possess: the quality of the ice is directly related to the quality of the water. Minerals and other more unpleasant constituents of the water will remain in the ice and, as a consequence, a coating may develop on the surface of the evaporator. This coating may be almost imperceptible to the naked eye, yet have a very strong influence on the operation of the machine. The coating will resist the action of the auger, tending to hold back the ice and slowing its progress toward the ice chute. The auger must then build up greater pressure, hence stress the bearings and overload the motor, in order to deliver ice. Rarely will chemicals alone remove this coating. Periodically, the auger should be removed from the machine and the walls of the evaporator should be rubbed down with a fine emery paper or steel wool along with the chemical application. In this way, the evaporator wall will be once again smooth and offer little resistance to the harvesting of the ice.

Remember, omitting preventive maintenance is guaranteed to reduce the effective life of an ice maker. On the other hand, overdoing this maintenance can also have an undesirable effect.

14

Typical Service Profile

It might be interesting to see what types of service problems you can expect to encounter over a period of time. I looked over a number of service calls, covering a two-year period, to see if any pattern emerged. In that two-year period, I had 296 service calls pertaining to commercial refrigeration alone. Of these calls, 146 (49%) were essentially concerned with the refrigeration system itself. A total of 128 (43%) dealt with the electrical system and the remaining 22 (8%) were what I would call mechanical problems. Let's take a closer look at each of these categories and get a little better feel of what you may run into.

REFRIGERATION

Into this category, I've placed all problems that involved taking temperature and/or pressure readings of the refrigeration system to verify the poor system performance described by the customer. A breakdown of refrigeration problems is as follows:

Problem	Number
Refrigerant Leak	87 (60%)
Defective TXV	18 (12%)
Dirty Condenser	16 (11%)
Freeze-up	8 (6%)
Miscellaneous	17 (11%)

LEAKS

As you can see by the numbers, leaks are a very common problem. Usually, a shortage of refrigerant is indicated by the fact that the unit is short cycling, i.e., cutting on and off rapidly. Other indications of this problem are bubbles in the sight glass, a warm suction line accompanied by a high superheat and, quite often, a cool condenser surface due to poor efficiency. Once you have determined that there is a shortage of refrigerant, you must find the source of the leak. If the system is low on refrigerant, there must be a leak or there must have been one at one time.

The common leak detecting tool is the halide torch. This detector will find just about any leak in a system. It is good for leaks as small as 2 oz. per year and, practically speaking, will suffice in more than 99% of the leak searches you will make. A handly helper with the halide torch is soap bubbles or some commercially produced equivalent of soap. Using the halide torch to get close, and the liquid soap to pinpoint the exact location of the leak, is a common procedure.

While there are many feet of tubing in most installations, the task of finding a leak is simplified by the fact that most leaks occur at only a few places. The first thing I do, in checking for a leak, is to visually eyeball the system for a sign of oil. Refrigerant leaking from a hole will disappear into the atmosphere, but it usually leaves a thin film of oil that does not evaporate. Any sign of oil leads me to check that point first. Failing to find a leak in that way, I check parts of the system that have proved to be typical sources of leaks over the years. The flare nut on the evaporator side of the TXV probably leaks more than any other point in the system. Flare fittings on driers and sight glasses also are popular spots. Usually, I will quickly scan all the flare fittings with the halide torch before going to the next step. On a number of occasions, the sight glass was leaking around the glass itself. Other areas worth giving a quick check are: all solder joints, the relief plug on a receiver, thin capillary tubes leading to controls, particularly at the point of entry into the control, and finally the evaporator coil itself, Figure 14-1.

Usually, a visual inspection will show whether an evaporator coil is a candidate for replacement. Evaporator coils that have been exposed to acid solutions, commonly found in salads, pickles, vinegar solutions, etc., show very distinctive signs of corrosion. The aluminum fins become quite brittle and break away when touched. The copper tubing itself may become discolored. When you see a corroded coil, it's a good bet that a leak is present. You may be able to repair the leak but, in so doing, may create another leak. It makes good sense to replace an evaporator that is in such bad shape, without even attempting to repair it. Should you succeed in your repair efforts, you will probably be quickly called back to repair still another leak. Only about 5% of leaks require the replacement of an evaporator coil. It's not a common problem but certainly a possibility.

Having checked all these points, there is almost a 100% probability that the leak has been found. If not, well, now it gets a bit troublesome. If the system is water-cooled, the condenser might have a leak into the water side. Removing the water line on the downstream side of the condenser, and again using the halide torch, may very well show a leak.

It only happened to me once, but a brand new roll of tubing had a pin hole in the middle of a long run. Of course, the run was concealed in a dropped ceiling and took literally hours to find. Eventually, you will find the leak, but it takes perserverance and, in some cases, luck. Sad as it is to say, on more than one occasion I recharged a system, not having found a leak, only to have it show up as a bigger, detectable leak, a year or two later. Practically speaking, there is a limit to the time you should spend trying to find a leak. If you can't find it in an hour or less, the chances are you won't find it at all. If the leak is that small, it's not worth

FIG. 14-1. The evaporator coil of this reach-in refrigerator has a leak in the return bend. Not a very easy leak to spot because of its inaccessible location.

it to you or the customer to invest more than an hour trying to find it.

That was a lot of words on leaks, but after all they take up 60% of the service calls.

TXV PROBLEMS

The commonest problem with the TXV is that it is either jammed open or jammed closed. If the compressor is cycling on and off on the pressure control, and the system appears to be adequately charged, chances are that the TXV is jammed closed. In most cases, the first step is to attempt to adjust the valve. Occassionally this is successful, with about one out of five valves, but more often than not a new valve is required. If, after opening the adjustment stem fully, and, after heating the sensing bulb in your hands, no refrigerant flows through the system, the valve should be replaced.

When the valve is jammed open, there is little or no superheat. Adjusting the stem, clockwise, should increase the superheat, if the valve is working. If superheat does not increase, and it usually climbs fairly rapidly in a properly functioning valve, say within 5 to 10 minutes, then again the valve should be replaced. You can see that the frequency of valve replacements is really not very high, so take your time and make sure you have thoroughly checked the other possibilities before condemning the valve. Remember the analysis presented earlier in the book before you go madly changing parts.

DIRTY CONDENSER

An air-cooled condenser with a heavy coating of dirt just can't dissipate the heat properly. Even a light coating of dirt will affect the refrigeration system, but usually not enough to draw the customer's attention to the problem. Dirt accumulation, that leads to a noticeable decrease in system effectiveness, is usually quite noticeable to the eye. It is standard procedure to check the condenser on all service calls, even when the problem is not caused by a dirty condenser.

A dirty condenser surface has a number of effects on the refrigeration system, some not immediately apparent. The most obvious is high head pressure, hence, decreased pumping by the compressor. This leads to warm temperatures in the refrigerated space. A longer term effect of high pressures is that the compressor has to work harder, hence, draw more current, resulting in an overheating of the compressor motor. This overheating may shorten the life of the compressor motor. Overheated motor windings may liberate moisture that will form acid, exist as free water and lead to a freeze-up, and/or attack the oil and decrease its lubricating capacity.

The little air that does pass through a dirty condenser will be rather hot. This hot air passes over the condenser fan motor, but does an inadequate job of cooling the motor. As a result, the condenser fan motor now overheats. Eventually, this will lead to motor failure when the motor windings short out or the motor bearings dry out. All this because of a little dirt.

Condensers that are located on the bottom of a piece of refrigeration equipment, Figure 14-2, are usually more clogged up with dirt than those located above the box. In any case, it is an easy matter to clean the condenser, if the dirt that has collected is dry dust and lint. Dry, caked dust can be removed with a stiff bristle brush and a blast of air or refrigerant will blow out the lint.

A tank of air, at a pressure of about 100 psi, is handy to have around for blasting out this dirt accumulation. A problem with an air tank is that the pressure drops quickly and, if you have too many condensers to clean, there may not be enough pressure to finish the job. I find that a blast of refrigerant works quite well and the pressure doesn't drop off as

FIG. 14-2. A relatively clean condenser. It is not unusual to be unable to see the condenser at all, especially when it is mounted close to the floor.

fast as with air. Of course, if you are in an area where an open flame is burning, you want to make sure that there is adequate ventilation. Nothing will clear out a kitchen faster than the clouds of phosgene gas created when your stream of refrigerant comes in contact with the stove flame.

Greasy condensers are a real problem. If they are mounted in a basement or on a roof, the problem is simplified. A high pressure stream of grease-cutting detergent, followed by a cleansing stream of water, will do a good job. When the condenser is self-contained, the problem is one of keeping the entire area from getting messy and of cleaning up after you're done. Frankly, I have never seen a neat way of doing this. A brush may very well get the surface grease, but the grease lodged in between the condenser fins can only be gotten out by a pressure spray. The best bet is to schedule the cleanup for a time when no one is around, so that you can make a mess and clean it up without disturbing anyone.

FREEZE-UP

Free water in a refrigeration system will accumulate at the TXV or cap tube and cause a blockage. This blockage can be mistaken for a defective TXV or clogged cap tube rather than being identified as a freeze-up. First, treat a blockage of this sort as a freeze-up and apply heat to attempt to remove the blockage. If the system does not respond, then you can assume that the flow control is defective.

Most freeze-ups are the result of a small amount of moisture in the system. They are readily corrected by the use of filter-driers or, in small systems, by the use of dehydrating fluids. Remember to make sure you are not voiding a warranty by using such fluids. My own experience has been good when using these fluids according to directions.

REFRIGERATION MISCELLANEOUS

On a couple of occasions I encountered capillary tubes that were clogged with something other than ice. There are tools around that enable you to remove the cap tube and clear the blockage by applying hydraulic pressure. I've never had occasion to do other than replace the tube. Many of the replacement cap tubes are provided in kit form, Figure 14-3. The kit enables you to size the cap tube for the unit you have. These all-purpose replacements work quite well if they are installed according to instructions.

It's important to note that you should not charge the system according to the nameplate if you have installed a new cap tube. Rather, use the slow, somewhat tedious method of observing the frost line and measuring superheat and head pressures, as described earlier.

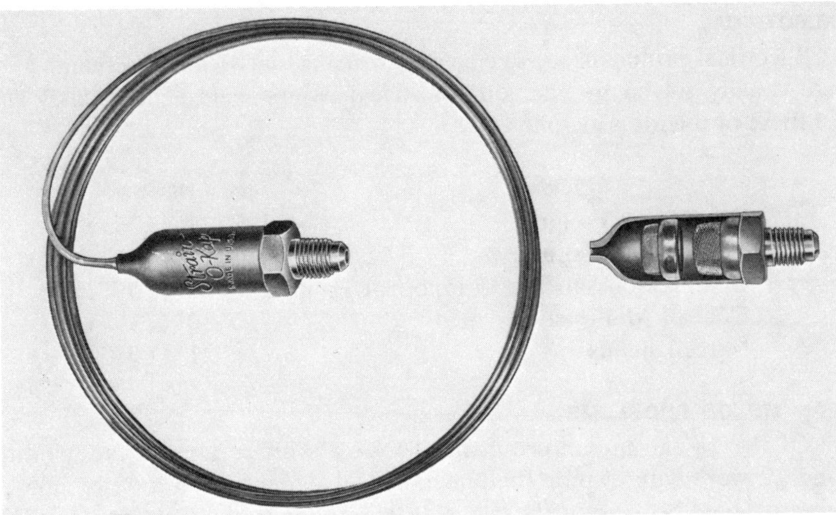

FIG. 14-3. Capillary tube designed for field replacement. By varying the length or by adding inserts, a small inventory will serve a wide variety of applications. *Courtesy Watsco, Inc.*

Oil level in a compressor is important. The only times I've had to add oil to a compressor was when oil had been lost through leaking shaft seals of open type units. The symptoms of an oil starved compressor are excessive noise, high running amperage, and possibly a slipping belt caused by the compressor seizing, or running tight, while the motor is running a full speed. Unless the compressor is pretty far gone, that is, has run for long periods with inadequate lubrication, the addition of the proper amount of oil will result in an immediate drop in running amps and a noticeable reduction in noise level. Full hermetic compressors should never need additional oil unless you've noted an obvious, measureable, loss of oil in the system.

Open-type compressors are invariably belt driven and the V-belt must be properly adjusted to assure that it does not slip. When properly adjusted, the belt will flex no more than about ½ inch when you apply fingertip pressure. Less flexing than this indicates that the belt is too tight. A tight belt will impose excessive pressure on the compressor seal and the motor bearing, and cause early failure of both. Too much flexing means that the belt is loose and points to a loss of efficiency and decreased cooling capacity. It is a good idea to check belts routinely with any service of a belt-driven unit. A belt that is dry or cracked, or one that is excessively worn, will eventually lead to failure. Regular replacement of drive belts is good preventive maintenance.

ELECTRICAL

Electrical problems are uncovered by using electrical instruments and are usually solved by changing electrical components. The breakdown of these problems is as follows:

Problem	Number
Fan Motor Failure	28 (22%)
Compressor Failure	23 (18%)
Compressor Starting Component Failure	26 (20%)
Control Malfunction	30 (23%)
Miscellaneous	21 (17%)

FAN MOTOR PROBLEMS

Of the 28 fan motor problems, 16 were with condenser fan motors and 12 were with evaporator fan motors. Condenser fan motors almost always failed because the bearings dried out due to overheating. In small commercial units, condenser fan motors have sealed bearings that cannot be lubricated. If properly cooled, they will run for years and years without a problem. The fact of the matter is that, at some time or another, the condenser is allowed to become dirty. This raises the air temperature over the motor and the bearings run hot. While there is an occassional electrical short, usually the motors will fail by seizing up. This creates a high current draw and eventually an electrical burnout of the motor.

Evaporator fan motors do not have problems with high temperature air. However, after many years of good service, the permanent lubrication eventually disappears and the motor seizes up. In many cases, a few drops of oil on the shaft, where it enters the bearing, can loosen the bearing and allow the motor to run again for a few more years. Most of the motors are replaced because the windings burned out or because the bearing was worn to the point where the shaft would not turn properly.

COMPRESSOR FAILURE

There may be a question as to whether or not compressor failure should be considered an electrical problem. Many, if not most, compressor failures are the result of a shortcoming in the refrigeration side of the system. Detection of the failure, however, is usually through electrical troubleshooting. Most compressors fail because either the run or start winding of the single phase motor is open or grounded. Your ohmmeter will enable you to determine that the compressor motor has failed.

In some cases the problem is a damaged valve within the compressor. This, of course, cannot be detected electrically.

A compressor that has failed electrically is said to be *burned out.*

Burnouts have been treated elsewhere and you might want to review that section at this time. There are a number of reasons for burnouts, only one being a defect in manufacture. More commonly, a shortage of refrigerant results in the compressor windings being improperly cooled. A dirty condenser will compel the compressor motor to work too hard, hence, to run hot. A malfunctioning starting component will put undue stress on the motor at start-up, again resulting in a burnout. A partial power failure or a drop in voltage (brown-out) will also cause excess current draw and speed compressor failure. A malfunctioning expansion valve or an overcharged cap tube system will allow liquid to get back to the compressor where it will either dilute the lubricating oil or actually get into the cylinder and slug the compressor. These are, as you can see, any number of conditions that will cause a burnout or mechanical failure. Each time a compressor is replaced, a determined effort should be made to track down the reason for the failure and to seek a cure for the primary problem.

COMPRESSOR STARTING COMPONENTS

The starting relay and the start capacitor are two major sources of electrical problems. They seldom fail due to inherent weakness, but because of deficiencies in the refrigeration system. The start capacitor provides the high torque necessary to get the compressor going. It is designed to provide about 20 starts per hour. Obviously, a properly functioning system should not have to start anywhere near 20 times per hour and, therefore, the capacitor should not be overtaxed. A system that is short of refrigerant, however, may very well cycle on and off as much as 20 times per hour, as will a system with a defective thermostat or pressure control that is improperly set. Under these conditions, the capacitor may very well break down and open or short out. When this happens, the compressor tries to start but cannot. This attempt to start results in high current draw and the overheating that goes with it. Result: compressor burnout. You can see that a properly functioning start capacitor is vital.

If you find that the compressor is trying to start, that is, it hums when energized and then cuts out on its overload, there is a good possibility that the capacitor is at fault. The sympton could indicate either a defective capacitor or a faulty starting relay. Of the 26 calls dealing with faulty starting components, about half were for relays and half for capacitors. Generally, it's a good practice to replace both, if one or the other is found to be inoperative. An inoperative capacitor puts great stress on the relay and the same can be said for an inoperative relay shortening the life of the capacitor.

CONTROL MALFUNCTION

Of the 30 control malfunctions, only 13 required replacement of the control, 5 being thermostats and 8 being pressure controls. In most cases, all that was required was an adjustment of the setting. In the case of pressure controls, the major complaint was high temperature in the conditioned space. Visual inspection usually showed that the evaporator was covered with frost and, in several cases, with a substantial layer of ice. This was attributed to the pressure control cut-in setting being too low, so that the self-defrosting feature of the refrigerator would not work. By increasing the setting by several psi, the box began to work properly and the problem was solved. In many applications, the pressure control is mounted on the condensing unit. In this position, it is subject to vibration which, on older controls, may result in the setting shifting. Sometimes the cut-out as well as the cut-in setting has changed, Figure 14-4.

Thermostats generally either work or don't work. While occasionally someone has *monkeyed* with the control and changed its setting, it was a more common to find the control to be completely inoperative. Usually, the sensing bulb or cap tube has lost its charge and, therefore, its ability to sense temperature change. The most common symptom of thermostate failure is that the refrigerator runs constantly, freezing the contents of the box.

FIG. 14-4. Vibration will often alter the cut-in and cut-out settings of this pressure control. Leaks sometime develop at the point where the sensing tube enters the bellows. *Courtesy Penn Div., Johnson Controls, Inc.*

A common problem in pressure controls is that the capillary, that connects the control to the point where the pressure is being sensed, has broken. This happens because vibration fatigues the tube itself, or more commonly, two tubes rub together and wear through. A cap tube rubbing against a refrigerant line or against another cap tube will eventually wear through. Many of these problems are handled as leaks, but occasionally the break cannot be repaired. If the leak is very close to the bellows, it may not be repaired without further damage to the control. In that case, a new bellows assembly, if available, can be installed. More commonly, the entire control is replaced.

ELECTRICAL MISCELLANEOUS

Many possibilities that happen rather infrequently can be considered miscellaneous. Broken wires or wires that are so old that their insulation has dried out and cracked off the conductor are examples. A good thing to remember is that the gage of the wire must not be reduced when you put in a replacement. This seems obvious, yet I've seen time and again where wires have been undersized by a serviceman. They work while he's looking at the equipment. Yet, over the long haul, they overheat and break down. When replacing wires, make sure the termination you use is tight and making good electrical contact. *Quick connect* terminals and *screw* terminals have to be tight to prevent arcing and burning of the replacement wire.

While compressor starting components are replaced rather frequently, the compressor overload rarely needs replacing. The failures that I have seen are of two types. Either the overload sensing disc (bimetal type) has burned out altogether, as a result of extended overload, or the calibration of the device has changed, resulting in nuisance trips when, in fact, no overload exists. This second malfunction is a bit tricky since the first impulse is to condemn the compressor when the overload trips. The snap-on ammeter will be useful here to actually measure the current when the overload opens. You might just find that it is at or below the nameplate rating of the compressor, in which case the overload is probably at fault.

A defrost time clock is designed to initiate the defrost cycle on a freezer and, at the appropriate time, terminate that cycle. A clock that either fails to initiate or fails to terminate the cycle must be adjusted or replaced. Usually, if there is a problem, the clock must be replaced. I've seen cases where defrost starts properly, but the clock never terminates the cycle. You can imagine the result, namely, a very warm freezer and the contents in a mess. Another type failure that I've seen is when the clock terminates the cycle, or at least turns the compressor back on again, but fails to de-energize the electric heaters. As a result, the box would not pull down in temperature. Instead of temperatures of about 0°F, the box might drop to 35°F and hold. Not only does this not freeze the contents of the box, but it puts an undue strain on the compressor and could result in a failure.

Most time clocks can be manually turned so that you can see what is happening at each stage of its operation. One problem you might note is that a defect that will appear when the clock is running normally may not appear when you are manually turning the timer. In that situation, about all you can do is invest the time necessary to watch the clock in normal operation.

All electrical equipment must receive voltage within ten per cent of the nameplate rating in order to operate properly. Occasionally, in period of *brown-out,* there may be a decrease that is significant enough to cause problems. The only practical thing to do is shut down the equipment until the *brown-out* is over. Usually, this is only an hour or two a day in really hot weather. If the low voltage situation persists and is a characteristic of the power distribution system in your locale, a *boost-buck* transformer might be useful. This will take, as an example, 208 volts and boost it up to 230 volts. They are relatively inexpensive, when compared to rewiring a building or bringing in new power lines. Incidentally, if you are in one of those rare areas with high voltage rather than low, the boost-buck transformer can be wired to drop the voltage.

Another typical voltage problem is not due to power company shortcomings, but to injudicious use of extension cords. Most extension cords are the light duty type meant for lights or small motors. Connecting such a small cord to a reach-in refrigerator, or worse yet a walk-in unit, will almost always produce a voltage drop through the line, not to mention high temperatures in the wire itself. Extension cords are just not good practice. They are something to be aware of when you're checking out erratically operating equipment.

The first thing to do with any problem is to see that the plug is in the wall. I've solved more problems than I care to think of by merely asking a phone-in customer to please go and once again, because they'll never admit they didn't, check to see if the plug is in the wall.

MECHANICAL PROBLEMS

A small minority of problems are what I refer to as Mechanical Problems. A leaking drain line in a walk-in refrigerator, a broken fan blade on a small condensing unit, or worn gaskets on refrigerator doors are examples of mechanical problems.

While mechanical problems are small in number, they can on occasion be a real headache. Local parts suppliers have a fine inventory of commonly used parts. When you need a gasket that fails every 20 years or a fan blade that never fails (except in the unit you are servicing) you may have a long wait or a long search before you can find the appropriate part. A word of caution: don't promise any delivery or job completion date on what seems to be an insignificant part. You may very well get hung up if you do.

ICE MACHINE SERVICE

Now, let's look at service calls on ice machines to see how they break down. Interestingly enough, only 10% of the ice machine service calls were directly involved with the refrigeration system. Dirty condenser coils, burned out motors, shortage of refrigerant, and defective TXV's were the most common problem.

The water circulating system ranked next on the list, accounting for 37% of the service calls. Typically, a water pump would have to be replaced or, in some cases, lubricated, a water solenoid valve had failed, a blocked tube needed clearing, or the entire water system had to be chemically cleaned. In several instances, there were water leaks that resulted in a partially formed batch of cubes.

Over half the problems (53%) were electrical in nature. Just about every control proved to be inoperative at one time or another. Improperly adjusted or defective bin thermostats were fairly common and the harvest cycle controls also came in for their share of misfortune.

The key to spotting just what is wrong is knowing how the machine is supposed to work and, when it does not work that way, to identify the device that controls the particular function. In addition, use your test instruments to verify, beyond a doubt, that the device you've identified as the culprit is indeed defective. I've suffered more embarrassment in ice machine service than with any other type of equipment. When you are in a hurry, it is easy to brand the wrong control as defective.

Know your machine, know how to use your instruments, and take your time and do it right.

TECHNICAL LITERATURE

In several places I've emphasized the importance of accumulating a library of reference material published by equipment manufacturers. Perhaps a word of caution is in order with regard to the use of this material. There is a tendency on the part of most people to worship the printed word. They feel that, if a bit of information has been put into print, it must be true. After all, who am I, a lowly serviceman, to question the printed word of a multi-million dollar manufacturer? Well, the truth of the matter is that a lowly serviceman can very often question that manufacturer and, upon occasion, find that the manufacturer doesn't know as much as he does.

I've run across wiring diagrams that did not accurately depict the manner in which a piece of equipment was wired. I've read service procedures that were downright unsafe. I've seen countless manuals that have at their beginning a number of *Errata* sheets, that is, corrections of mistakes that were printed in the original publication. The fact is that the people who put together technical information are human and subject to error, omission, hasty conclusions, and even deficient technical background. Fortunately, I can also state that most of the material I've received from manufacturers has been adequate, at least, and extremely enlightening, at best.

Be aware that all that is written by a manufacturer may not be gospel. You must develop confidence in your own ability to examine a service situation on the spot and, using the available literature as a guide, solve the problem. The solution may fly in the face of all that is written but, in spite of that, you must have the confidence to go ahead and find that solution.

After you have broken this *new ground,* it might be fun to communicate with the manufacturer and get his reaction. When I sat on the manufacturer's side of the table as Director of Quality Control of a heating/air conditioning company, I was more than delighted to hear from the field. It showed that people were reading our technical manuals. If we made a mistake now and then, well, it happens. To have a mistake go on for years, and not be uncovered by technical people in the field, is the real shame. On any number of occasions I've called manufacturers with questions about their service literature and invariably I got straight information and thanks for bringing a potential problem to the attention of the company. Incidentally, nobody likes a smart ass. Do it in a tactful way. You may be wrong and the *come down* is a lot easier if you have broached the question tactfully.

LISTENING

The commercial refrigeration repairman is faced with many more challenges than, say, the air conditioning serviceman. The variety of equipment is quite extensive and, as a result, the possible problems are also quite extensive. This is mentioned not to discourage the potential commercial service technician but rather to motivate him by pointing out that "Variety is the spice of life."

In spite of the variety encountered so far, there is a common thread running through all the equipment described. They all have an evaporator, a compressor, a condenser and a flow control device of some type. They all depend upon the principles of the mechanical vapor cycle, and they can, therefore, all be diagnosed using the procedures we've presented. In addition to variety then, we also have the reassurance that "There is nothing new under the sun."

A final word or two concerning your approach to commercial equipment. It's a good idea to ask the customer what the problem is. He may not know how to describe it in language that is technically precise, but he does know he has a problem and, in most cases, he will be able to give you a head start in isolating his problem. A really good service technician knows how to communicate and knows that a major part of communication is *listening*. You are not there to impress the customer with how smart you are, but rather to determine what his problem is and to solve it.

SUMMARY

This little look at what a typical commercial serviceman may encounter in the *real* world should bring a few things into focus. *First,* while

FIG. 14-5. Remember: a neat service vehicle is a happy service vehicle. But it sure is hard to keep it that way.

being able to detect leaks and add refrigerant to a system may be an important part of the total picture, it is by no means all of it. If that's all you know, you had better buckle down and learn a bit more. *Second,* there are enough problems that require a working knowledge of the refrigeration cycle to warrant your studying and applying the Mollier diagram in order to understand everything that can go wrong in the system. *Third,* if you don't know some electrical theory and if you can't use a voltmeter, ohmmeter, and ammeter, you have no future as a commercial refrigeration serviceman. *Finally,* the skills required of a good refrigeration mechanic are as demanding as, if not more demanding than, any other technology and the really good service technician can hold his head high for his contribution to society.

Index